WIRELESS BROADBAND HANDBOOK

Wireless Broadband Handbook

Regis J. (Bud) Bates Jr.

McGraw-Hill
New York   Chicago   San Francisco
Lisbon   London   Madrid   Mexico City
Milan   New Delhi   San Juan   Seoul
Singapore   Sydney   Toronto

McGraw-Hill

*A Division of The **McGraw·Hill** Companies*

1 2 3 4 5 6 7 8 9 0   DOC/DOC   0 9 8 7 6 5 4 3 2 1

ISBN 0-07-137161-3

The sponsoring editor for this book was Steve Chapman and the production supervisor was Sherri Souffrance. It was set in Century Schoolbook by MacAllister Publishing Services, LLC.

Printed and bound by R. R. Donnelley & Sons Company.

CONTENTS

About the Author   xv
Acknowledgements   xvii

Chapter 1   The History of Wireless   1

                  Wireless Radio Systems   3
                      Free-Space Communication   5
                      Frequency Spectrum   7
                      The Modulation Process   9
                      Radio Propagation   11
                      Microwave Repeater Systems   15
                      Satellite Radio Communications   17
                  Standards Groups   19
                      ANSI   19
                      TIA   19
                      ECSA   21
                  Spectrum Usage   22
                      Regulatory Control   22
                      Cellular Service   22
                      Analog Cellular (AMPS and TACS)   24
                      Digital Cellular   24
                      Personal Communications Services (PCS)   25
                      Universal Mobile Telephone Systems   25
                  Radio Channels   25
                  Multiplexing Arrangements   28
                      Frequency Division Multiple Access (FDMA)   28
                      Time Division Multiple Access (TDMA)   29
                      Code Division Multiple Access (CDMA)   30
                  Modulation Techniques Used   32
                      Modulated Signal Envelope   33
                      Amplitude Modulation   34
                      Frequency Modulation   35
                      Digital Modulation   36
                  Light-Based Systems   37

Chapter 2   Radio Technologies and Systems   39

                  Wireless Methodologies   40
                  Radio Frequency Characteristics   40
                      General Aspects   40

Microwave Communications   49
What About Bandwidth?   53
Satellite Communications   56
   Commercial Providers   57
   How Do Satellites Work?   57
   Satellite Frequency Bands   57
   Orbital Slots   61
   Communications   61
   LEO versus GEO   64
   Niches in the GEO Sphere   64
   LEO Meets GEO   65
   Space Security Unit   65
   The Market for the Network   65
Satellite Characteristics   67
   Latency   68
   Noise   69
   Bandwidth   69
   Advantages   70
Low-Earth Orbit Satellites (LEOs)   70
   Low-Earth Orbit   71
The Benefits of These Service Offerings   74
Global Positioning System   77
   Direct Broadcast Satellites   77
   Communication Satellites   77
Other Commercial Applications   78

Chapter 3   Access Techniques for Radio-Based Systems   81

Frequency Division Multiple Access (FDMA)   82
Digital Cellular Evolution   84
Time Division Multiple Access (TDMA)   84
   IS-136 TDMA   86
   Capabilities Provided Through TDMA   87
   Voice Quality Improvements with IS-136 TDMA   89
   Data Services Currently Available with TDMA   90
   Automatic and Dynamic Channel Assignment for IS-136   91
   Microcellular Evolution for IS-136   92
   Future High-Speed Packet Data Wireless Access Using EDGE   92
   Additional Considerations for 136 HS   93
   Extended-TDMA   94

Contents

Code Division Multiple Access (CDMA)   95
    The CDMA Cellular Standard   97
    Spread Spectrum Goals   97
    Spread Spectrum Services   99
    Security   100
    Synchronization   100
    Balancing the Systems   100
Common Air Interfaces   101
    The Forward Channel   101
    Overhead Channels   102
    The Reverse Channel   103
    Traffic Channel   105

Chapter 4   Cellular Communications   107

Why Digital?   111
Coverage Areas   112
Analog Cellular Systems   113
Log On   114
Monitoring Control Channels   115
Failing Signal   115
Setup of a Call   116
Setup an Incoming Call   117
Handoff   117
    Setting up the Handoff   118
    The Handoff Occurs   118
    Completion of the Handoff   118
The Cell Site (Base Station)   119
The Mobile Telephone Switching Office (MTSO)   119
Frequency Reuse Plans and Cell Patterns   120
    Overlapping Coverage   122
    Cell Site Configurations   122
    Sectorized Cell Coverage   123
    Tiered Sites   124
Reuse of Frequencies   124
Allocation of Frequencies   125
Establishing a Call from a Landline to a Mobile   126
Intersystem Handoff   128
    Handoff Completion   128
    Hand Back   129
    Handoff to a Third MSC   129

	Seamless Networking with IS-41 and SS7	130
	Automatic Roaming	131
Chapter 5	Personal Communications	133
	Current Cellular Standards	136
	FDMA	136
	TDMA	136
	Digital Systems	137
	Digital Cellular Evolution	138
	Time Division Multiple Access (TDMA)	139
	Penetration	140
	The CDMA Cellular Standard	142
	CDMA Development Group	142
	CDMA-PCS	142
	More Sophisticated Vocoders	144
	Capacity Improvements	145
	CDMA Benefits	145
	CDMA Today	147
	Rationale Behind CDMA's Popularity	148
	Soft versus Hard Handoff	148
	Over-the-Air Activation	149
	What About Data?	150
	Circuit Mode Asynchronous Data/Fax Rates	151
	Simultaneous Voice and Data	151
	Packet Data Services	152
	PCS Providers	152
Chapter 6	Global System for Mobile (GSM)	155
	Change Is Underway	158
	GSM Concept and Services	158
	The GSM Architecture	159
	The Air and Link Interfaces	161
	The Access Techniques Used	162
	Traffic Channel Capacities	163
	Control Functions	164
	The Data Burst	164
	Speech Coding Formats	165
	The Network Structured Protocols and Interfaces	165
	Some Thoughts on GSM	167
	The Need for Interoperability	168

Contents

	Network Interoperability	170
	Enhanced Voice Services	170
	Add-On Technologies — iDEN™	173
	Improved Spectral Efficiency	173
	Motorola's VSELP—Coding Signals for Efficient Transmission	174
	QAM Modulation	175
	Multiplied Channel Capacity	175
	The Advantage of Integration	175
	The Control Channel (CC)	177
	Service Areas and Licensing Blocks	178
	Innovation and Integration	179
	Spectral Efficiency with Frequency Hopping	179
	Digital Transition	180
Chapter 7	Wireless Data Communications Services	183
	The Wireless Revolution	184
	Voice to Data	185
	The Wireless Data Market	185
	Wireless Data and the Spectrum	187
	Spectrum Regulation	187
	Unlicensed Spectrum	188
	Wireless Data Transmission:How It Works	188
	Session versus Packet Transmission	190
	Cellular Digital Packet Data (CDPD)	195
	Circuit-Switched Cellular Digital Packet Data (CS-CDPD)	195
	Packet-Switched CDPD	195
	Packet Data Communications Are More Efficient	197
	Wireless Application Protocol (WAP)	199
	SMS	201
	National SMS Interworking	201
	Person-to-Person Messaging	202
	Voice and Fax Mail Notifications	203
	Internet E-mail Alerts	203
	The Wireless Internet	203
	General Packet Radio Systems	205
	GPRS System Architecture	206
	Bearer Services and Supplementary Services	209
	Simultaneous Usage of Packet-Switched and Circuit-Switched Services	210
	EDGE — The Next Step in Wireless Data	210

GERAN   211

UMTS   211

   UMTS Access Network (UTRAN)   212

The Wireless Data Industry   214

   System Integrators   214

   Software   214

   Hardware Developers   215

   Carriers   215

Wireless Data: Apparatus Types   216

Chapter 8   Wireless Local Area Networks (WLANs)   219

Wireless Local Area Networks   220

Defining the Wireless LAN   222

Applications for Wireless LANs   224

Benefits of WLANs   225

How WLANs Work   227

WLAN Configurations   228

   Independent WLANs   228

   Extended WLANs   229

   Infrastructure WLANs   229

WLAN Technology Options   230

   Spread Spectrum   230

WLAN Customer Considerations   231

   Range/Coverage   231

   Throughput   231

   Integrity and Reliability   231

802.11 Specifications   232

IEEE 802.11 Architectures   233

   IEEE 802.11 Layers   233

   Physical Signals   234

   Timing Is Everything   235

   Clear to Send?   236

   Roaming   237

   MAC Layer and Data Payload   238

Home Networking   238

802.11b versus HomeRF Networks   238

   Wireless Access Point   241

   Wireless Adapter   242

   Realities of Wireless   242

Some Motivation   243

   Mobile IP   244

Contents

Faster Wireless Standards: 802.11a 244
Frequencies for All 245
Bran Anyone? 247
Taming the Standards Beast 247
Interoperability Problems 248
Dental Hygiene Anyone? 248
What Is Bluetooth? 248
   Bluetooth Roots 249
   Compliance 250
   Voice 251
   Data 251

Chapter 9   Wireless Innovations in Broadband 255

Wireless Innovations 256
The Market in General 256
From Suitcase to Palmtops 257
The Ricochet Network Architecture 260
   The Components 260
Radio Frequency Spectrum 263
   Frequency Hopping, Spread Spectrum Technology 263
   Packet-Switched Networking 264
Middleware, Custom Protocols, and Proxies 264
Mobile IP 267
TCP/IP over Satellite 268
Satellite and ATM 270
Charting the Rules for the Internet 270
   Tailoring IP Can Accelerate Throughput 271
Teledesic Technology Overview 273
   Seamless Compatibility 273
The Teledesic Network 273
   Fast Packet Switching 275
   The Satellite Constellation 276
   Multiple Access 277
   Network Capacity 278
Wireless Local Loop (WLL) 279
WLL Technology Shakeout 280
Architecture of a Wireless Downstream System 280
Frequency Bands and Limitations 281
Receiving the Signal at the Subscriber 281
Wireless Local Loop (WLL) 285
   Not for Everyone 286

What of the Bandwidth?   289
Enter Local Multipoint Distribution Services (LMDS)   290
The Reasoning Behind LMDS   290
Network Architectures Available to the Carriers   293
TCP/IP over LMDS   294

Chapter 10   Emerging Wireless Standards   297

Wireless Standards   298
GPRS   299
EDGE   304
  What Is Special About EDGE?   305
UMTS   306
Mobile Internet — A Way of Life   307
Applications of the Wireless Internet   311
Visions of Wireless   312
Positioning the Mobile Industry   315
Key Technologies   316
  UTRA   316
  Multi-Mode Second Generation/UMTS Terminals   316
  Satellite Systems   317
  USIM Cards/Smart cards   317
  Internet Protocol (IP) Compatibility   318
Spectrum for UMTS   319
The cdma2000 Family of Standards   319
  Purpose   319

Chapter 11   Wireless Applications   323

Using Wireless   324
There Is a Bug in My Soup   325
Wireless Intenet Starts It Off   326
Applications and Features   327
TV as an Application   331
What About Dick Tracy?   332
Web Through the Sky   333
Through the Air with Non-LOS   337
Medical Prescriptions   338
Dental Floss and PDAs   340
SOHO Long, It's Been Good to NOYO   342
Students and Profs Come Together   343
Shopping Panorama   345

Contents

In Your Face   346
Ad Hoc Meetings with InfraRed   348
Finally the Set Changes   349
Phoneless in Chicago   350
Where Am I Going?   351
How Do I Get There from Here?   351
Final Comments   355

Acronym List   357
Glossary   367
Index   387

ABOUT THE AUTHOR

Regis J. (Bud) Bates Jr.
President
TC International Consulting, Inc.
PO Box 51108
Phoenix, AZ 85076-1108
Tel. (800) 322-2202
Fax (800) 260-6440
http://www.tcic.com

Mr. Bates has more than 35 years of experience in telecommunications and information systems. He oversees the overall operation of *TC International Consulting, Inc.* (TCIC) of Phoenix, Arizona. TCIC is a full-service management consulting organization that specializes in designing and integrating information technologies. TC International Consulting leads the pack in strategic development and implementation of new technologies for carriers and corporations alike.

Bud's experience has served in major network designs from *local area networks* (LANs) to *wide area networks* (WANs) using high-quality, all-digital transmission services: T1, T3, and SONET/SDH. His studies and recommendations have resulted in significant financial savings. One project included the design and implementation of a Frame Relay network that spanned over 14 countries and 80 locations. This project resulted in huge monthly savings while preserving sub-second response times across the network.

His articles have been published in Network World, Information Week, International Journal of Information Management, and others. He has authored numerous books published by McGraw-Hill and Artech House. His recent published books Voice and Data Communications, Third Edition and the Broadband Telecommunications Handbook are on McGraw-Hill's 2000 bestseller list. Bud also develops and conducts various public seminars throughout the world, ranging from managerial overviews to very technical instructions on voice, data, and LAN communications. He spends much of his time working with the major telecommunications manufacturers in training their staff members on the innovations of technology and the convergence of voice and data networks for the future. Many of his materials are used throughout the higher education institutions in certification and graduate-level classes in telecommunications management.

Mr. Bates holds a degree in business management from Stonehill College, Easton, Massachusetts. He has completed graduate-level courses at Lehigh University and Saint Joseph's University, specifically in financial management and advanced mathematics.

MEMBER

NATIONAL
SPEAKERS
ASSOCIATION

ACKNOWLEDGMENTS

Well, here I am again, staring at a pile of papers and graphics that have accumulated over the past couple of months. What it will soon become is the latest in a book series that I have embarked on with McGraw-Hill. This book deals with the issues of wireless communications and broadband communications as the two converge into the world of 3G. As always, I chose to write the way I think, that is, in a non-technical fashion. I also think in terms of what a CEO, CFO, or CIO needs to know about the technologies we will discuss herein. As I usually start off, I try to answer the three management questions:

What is it?
What will it do for me?
What is it going to cost me?

But before I go too far into the plot, I owe a lot of credits to many people. Some of the people I had to deal with either on a regular or irregular basis. First, I owe another debt of gratitude to Steve Chapman, McGraw-Hill's executive editor on this book. Steve is a very patient person. He has had to occasionally prod me to keep going, even though he knew the pressures I had in other areas. He is a true gentleman whose efforts I appreciate while he has to go to bat for me. In addition to Steve, there are numerous other people who aided in the editing and production of the book, too many to name. Special mention is given though to Molly McHugh who had the dubious pleasure of trying to cajole the edits and rewrites out of me on a regular basis. Her tenacity is what got the final document together.

Two people in my office deserve the lion's share of the credits for the ultimate graphical representation of this book. First is Gabriele (many of you have seen her name in other books). Gabriele is my wife and partner of more than 30 years. She provides the constant support at home and the office when I just can't seem to get the motivation to work nights and weekends on the production of a book. This is especially true when I have been on the road for days on-end and have numerous other tasks tugging on my time. Gabriele really deserves to have her name on the cover of this book.

Secondly is a young lady with whom I am most impressed by her energy and enthusiasm. Amber Hartmann is a graphic artist who takes many of the figures I create in stick drawings and turns them into art that we can actually use. Amber's age belies her maturity and ability to jump right in and get things done quickly, even if I change them constantly.

With this team behind me, I had no choice but to succeed. I owe each of these people special thanks and appreciate their undying loyalty and support to me. Anyone could succeed at writing a book with the support I had behind me.

Additionally, there are a number of vendors with whom I continually converse or gather information. I thank them all collectively because there is not enough room allotted for individualized thank-you notes.

Finally, I have one large group that I sincerely appreciate; that being you, the reading public! Without you and your interest in what I have to say, I would not be in this position today. I receive many calls and e-mails from readers who just want to let me know that they enjoyed my opinion or the way I present an idea. I hope I can continue to win your support.

My best wishes to you all!

The History
of Wireless

Welcome to the world of wireless communications and the logical extension to the broadband architectures that are emerging as the future of the industry. No aspect of communications will be untouched by the wireless interfaces; no part of our working environment will be left untouched either. As the world changes and the newer technologies emerge, we can expect to see more in the line of untethered communications than in the wired interfaces. Users are becoming more mobile; data is being retrofitted to accommodate a wireless and a mobile user; and finally the new protocols for handheld devices (*Personal Digital Assistants* [PDA], data-equipped cell phones, Internet-enabled devices) are all being adjusted to accommodate the higher speeds demanded by the end user.

With this in mind, we shall venture into a new world of wireless local loop, wireless handheld devices, wireless telephony, wireless CATV access, and wireless video-conferencing systems that will all be geared to meet the demands of this form of voice, data, video, and multimedia application.

When I agreed to write this book for the publisher, I was drawn back to a book that I published back in the early 1990s, called *Wireless Networked Communications*. The reason for the flashback is that I was ahead of the industry and myself at the time. The book discussed the many aspects of voice telephony over wireless communications and minimal data aspects. However, the topic of wireless data and LAN services is now very intriguing to many people. Moreover, the use of wireless communications for instant Internet access and messaging services becomes the icing on the cake.

This book will begin covering the topics associated with wireless communications. From there, we will take the various components of the wireless applications and build upon them to be sure that the many subjects are covered. In most cases, our discussion topics will stay on the non-engineering side of the business. I do not want to lose a valued reader nor do I wish to insult anyone's intelligence. This book is written for the non-radio (RF) person, the non-engineer, and the non-geek! Each of these persons can find many topics written in the mode that better satisfies their engineering curiosity. However, some of the pieces of this book will appeal to that person because the topic must address the applications, which is what the engineers must develop products to satisfy. Without an application, the technology must search for one. This means that the technology will not be as readily accepted without a "killer" application. Keep that in mind as you read this content.

For the non-engineer, this book will attempt to take each of the wireless components and break them down into pieces that are more manageable, more understandable, and more application specific. You can learn a lot from this book if you hang in there and stay with it. No author can profess to address all audiences with full acceptance and knowledge. Therefore, you can expect that this book will do its best to cover the bases with due diligence on the following:

- Radio-based and light-based systems
- Satellite-based systems
- Microwave systems
- Cellular communications
- *Personal Communications Services* (PCS)
- *Global Services for Mobile* (GSM)
- Data over wireless
- Wireless *local area network* (LAN) (802.11b)
- Wireless *wide area network* (WAN)
- 3G wireless
- Applications

Wireless Radio Systems

Probably one of the most interesting technologies in the industry today is the wireless world. This statement is not founded on the latest and greatest in communications breakthroughs. All too often, we hear how evolving technology will unshackle us from the traditional pairs of wires that provide our present-day communications. Wireless, however, has been around for decades in a variety of uses and techniques. Recently, newer applications breathed life back into mundane services, sparking international interest in the applications, bandwidths, and legalities of wireless communications. Today, everyone is buzzing about the future of the wireless world and our abilities to communicate in general.

Wireless communication is nothing new. Since the early days of civilization, various forms of communication took place without the advantage of physical connectivity. In tribal jungle environments, drums were a primary means of communicating. As message senders beat on either drums or hollowed-out logs, the reverberating sounds were interpreted at the other end. In many situations, the drumbeat would travel only a limited distance, so various relay points were needed. The first receiver would acknowledge the sender's message through a series of return drumbeats, then relay the same message to the next receiver. This limited form of transmission met the need, but was subject to a lot of noisy interference and misunderstanding. Therefore, the message was sent repeatedly to minimize potential errors in interpretation. It was crude, but it worked.

In early American times, Native American tribes used smoke signals as a limited-distance form of communication. Drawbacks to smoke signals included distance limitations based on *line of sight* (LOS), a limited alphabet, and errors caused by the wind. If the smoke puff was blown away or dissipated too soon, the communication was lost.

The introduction of the semaphore flag deepened the scope of communication through an enhanced alphabet that could address the language needs, but was also limited to line-of-sight daytime operation. Hardly a reliable or widely available capability, it was effective in certain circumstances. Because semaphore signaling allowed a full alphabet to be used, messages could be more extensive and detailed. Drawbacks were the limited distance (requiring constant relay of the information), the added time needed to send detailed messages, and the risk of an undetected interception of the message.

In the nineteenth century, light beams were used for short-haul communications, particularly in military contexts. Very detailed messages could be transmitted by a coded sequence (Morse code) of blinking lights from sender to receiver. Again, this was effective over limited distances and provided a quiet, yet visible means of communication. Drawbacks included limited distance, unauthorized reception of information due to visibility at various angles, and risk of interception. Security was always suspicious, so a form of alphabetic encryption was introduced as a safeguard. This required an ever-changing code set, along with special handling and extra time to manually decipher the transmitted message. Furthermore, the cipher code had to be kept current at all locations to achieve correctness.

In radio transmission, human speech must first be converted to an electrical signal. This signal is analogous to the composition of the sound-pressure changes produced by the human voice; hence, the term *analog communications*. The analysis of sound waves is a key part of radio communications theory. Knowledge of radio principles is critical to understanding how the various wireless communications techniques function. In early telecommunications systems (particularly the telephony world), radio was an integral part of network development. As newer systems emerged, modifications and enhancements enabled networks to carry all forms of communication, including voice, data, telegraph, image, fax, and video.

Free-Space Communication

Radio systems propagate information in free space. This free-space communication obviates some of the problems faced by other transmission systems. For example, wired systems require a physical medium and are difficult to install in certain geographic areas. Advantages of the radio system include the capability to

- Span bodies of water, such as lakes or rivers, where a cable facility would require special treatment to prevent seepage onto the copper conductors
- Overcome transmission obstacles posed by mountains and deep valleys, where cable costs would be prohibitive to install and difficult to maintain
- Bypass the basic interconnection to the local telephone provider (Telco) or *post telephone and telegraph* (PTT) company

In Figure 1-1, the local service provider realized that using a cable-based system would require an underwater run of wiring that would be prone to water leakage. An alternative was an over water (aerial) pull requiring extensive cable support systems (including pole construction with guyed wires) that would be prone to wind destruction and cable deterioration.

Figure 1-2 depicts the use of radio systems to overcome the terrain problems associated with a mountain or valley. Rather than construct a cable system rising over the mountain, the local provider used a radio system as a relay point.

Figure 1-3 shows a new application for radio-based systems; it could be used as a means of bypassing the normal cable route from the local supplier.

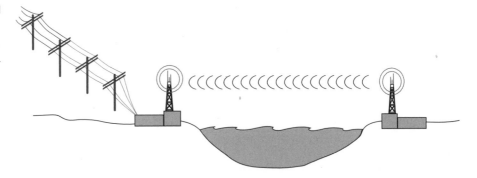

Figure 1-1
Radio in lieu of an aerial pull over water

Figure 1-2
Terrain problems can
be overcome.

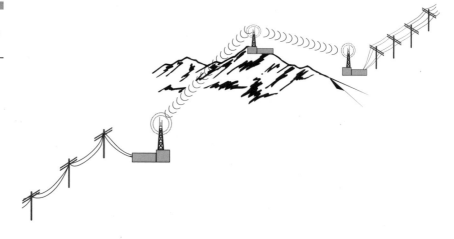

Figure 1-3
Private connections
to bypass the
Incumbent Local
Exchange Carrier
(ILEC)

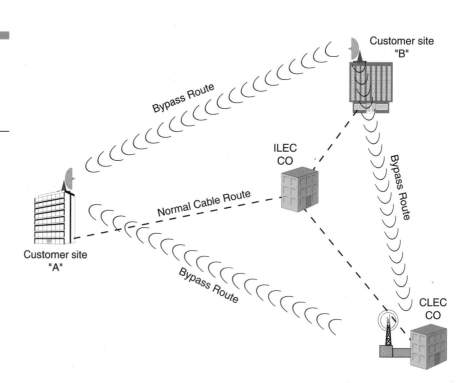

The dotted line represents how a leased-line facility would run from customer location A through the local supplier to customer location B. These leased lines would bear monthly or annual rental costs from the local supplier. Customers could avoid the high rental costs by using a private (owned or leased) radio system.

The use of radio waves in free space requires the conversion of the human voice (or other form of information) from sound to an electrical equivalent. The equipment components used to perform this conversion operate similarly in every radio system. Although different names may be used to describe the components, the functionality is important.

Frequency Spectrum

Radio systems such as two-way cellular, personal communications, microwave, and satellite all operate on a single frequency. Some use different frequencies (two) for the transmission and reception (send and receive) of information. However, they operate on a specific frequency, leaving no variations. In the U.S., these systems are licensed through the *Federal Communications Commission* (FCC) frequency registration program and have virtually no flexibility. If the frequency you are designated to operate on is busy, you wait.

Radio-specific frequencies can be integrated into a network easily. Because it is fixed function and frequency functional, a specific frequency is very predictable. Many of these systems have a range of interface capabilities; however, the primary interface system is the use of a multiplexer through a punch block (66 Block), which implies that the system is wired to wireless.

The radio system broadcasts its information out from a transmitter to receiver based on a fixed frequency. The energy is spread over a very limited spectrum; this concentration is therefore easily received and detected in the frequency spectrum. Two-way radio, PCS, and cellular systems fall into this same domain.

The operations are a function of both the length of the wave and the frequency in which the wave is produced. If we want to carry many voice calls, we need the bandwidth. The bandwidth will occur on the average number of cycles per second available to place the voice onto the radio wave. The more waves that are in a one-second period of time (the frequency), the more information we can carry. Figure 1-4 describes the frequencies and the length of the wave for the various radio and light-based systems; the shorter the wave, the higher the frequency.

Look at the short-wave radio frequency. It states that a wave is 10,000 (10^4) meters long. This means that 10,000 waves (10^4) per second using a 10,000 meter length can carry a very limited number of channels, whereas in microwave radio frequencies, the wavelength is much shorter (10^{-2}), but there are 10 billion (10^{10}) waves per second. We can certainly carry many more conversations on these radio channels.

Sound has two constantly changing variables: amplitude (the height of the signal) and frequency (the variable rate of change in a specific period of time). The pattern is normally represented by a sinusoidal waveform, as shown in Figure 1-4. The waveform representing the electrical equivalent of human speech is a function of both the amplitude (or value of current) and the frequency with respect to time. A complete cycle of the waveform is represented at the starting point (A) through the 360-degree cycle to the ending point (E). The complete cycle that occurs in a one-second time frame is called a hertz (Hz); 1 hertz is one cycle per second. The number of cycles that occur in a one-second period is the frequency. The frequency of standard speech is represented at 3,000 cycles per second, or 3 kilohertz (kHz). Human speech, therefore, is converted into a 3-kHz waveform, which can then be modulated onto a radio-based carrier.

Because the waveform represents the electrical equivalent for analog transmission, it will also have a certain velocity with respect to distance and time. In free-space radio transmission, the electromagnetic wave moves through the air at the speed of light (186,000 miles per second). Radio

Figure 1-4
The wave

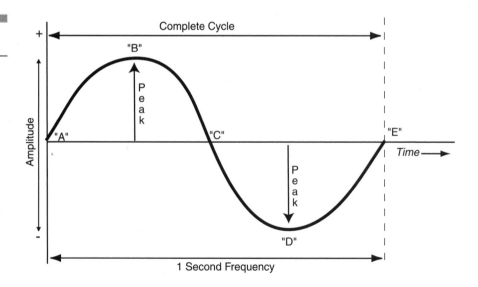

waves can be produced and transmitted across a wide range of frequencies, starting at about 10 kHz up through the millions of hertz (megahertz, or MHz), and even billions of hertz (gigahertz, or GHz). Figure 1-5 shows the typical frequency spectrum for the transmission of various forms of energy. The spectrum ranges from radio frequencies through light frequencies and finally to X-rays and cosmic rays.

The Modulation Process

Once a determination has been made as to what frequency will transmit the information, the information must be modulated (applied) to a carrier that operates in this frequency range. Human speech, as the example being used, generates a range of frequencies of approximately 100 to 5,000 Hz. However, most of the usable and understandable information from human speech is contained within the 3,000 Hz range. Therefore, as an economic expedient, commercial transmission systems are manufactured to deliver a range of frequencies (bandwidth) limited to the 3-kHz range. This is accomplished by breaking down the frequency spectrum into 4-kHz slices, whereby the electrical wave is applied to the carrier wave. However,

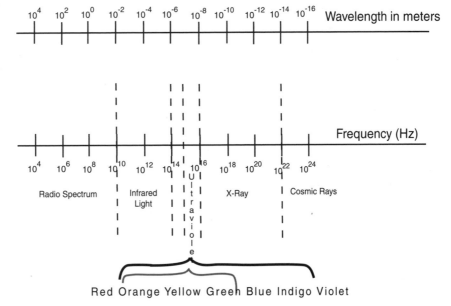

Figure 1-5
Frequency spectrum for the various forms of energy

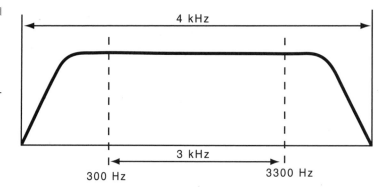

Figure 1-6
Bandwidth of a voice
grade channel for
narrowband
communications

bandpass filters are installed to transmit only those frequencies in the 300- to 3,400-Hz range. In Figure 1-6, the bandwidth is broken down into the 4-kHz channel, with the bandpass filters installed to allow only 3 kHz of frequency use.

Because 10 kHz is the usual starting point of radio transmission, it would be unreasonable to try to transmit speech information below 10 kHz in the radio spectrum. Therefore, a higher frequency range is used, whereby speech can be modulated onto a carrier signal at the transmitting end. As the speech is introduced to a piece of equipment—in this case, a modulator —it is applied onto a carrier frequency to change the actual wave by producing an envelope of the information. The modulated wave represents the sum of the carrier wave plus the signal to be transmitted. Modulation will be addressed later in this chapter.

Although the signal is carried in free space (the airwaves), the following factors must be considered in relation to the type of radio transmission system used:

- A decision must be made on whether to use line-of-sight, point-to-point, or omnidirectional transference via broadcast communications.

- Noise will always be a factor because its presence will degrade the signal.

- Power output will directly affect the distance that the signal will travel.

- Loss or attenuation will be a factor because a radio signal will be diminished as it passes through certain insulating materials and will experience gain when it passes through conductive materials or reflects off other objects.

- Heavy rain or snow will absorb some of the transmitted signal in certain frequency bands (such as, microwave and satellite).

Although this list is not all-inclusive, the most prominent factors are generic for most radio-based systems. Each factor must be given serious consideration prior to setting expectations for performance.

Radio Propagation

Depending on the band selected, the characteristics of propagation will vary. In general, when the signal is transmitted through an antenna device, the signal will travel along the earth's curvature (see Figure 1-7). As the signal emanates in all directions (or in a point-to-point direction), the energy follows the earth's curvature. In some cases, reflected power off the earth's surface helps achieve the desired result. At lower frequencies (very low, low, and medium bands), the signal follows the curve of the earth's surface in what is typically called a ground wave. The distance that the wave travels is a function of the amount of power generated by the transmitting device. Power output is selected to cover specific distances and areas.

At the *high frequency* (HF) band, the ground wave is absorbed and attenuated very quickly. However, the radiated energy also has an upward movement in which the signal reaches approximately 40 to 300 miles above the earth, entering the ionosphere. In the ionosphere, the radio waves are

Figure 1-7
Signal travels along the curvature of the earth.

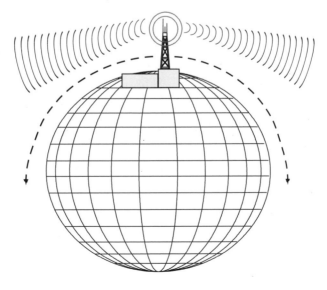

Figure 1-8
High-frequency
ground wave

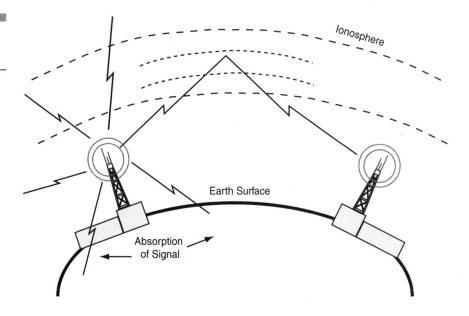

Figure 1-8
High-frequency
ground wave

refracted at various angles and bounced back down to the earth. This type of transmission enables radio signals to be directionalized and transmitted at much lower power output. Figure 1-8 illustrates this form of high-frequency transmission.

At the *very high frequency* (VHF) band, the signal is transmitted in straight lines. A directional antenna can be used to direct the signal in a LOS path. A certain portion of the signal can be reflected off the ground along this same path and get to the same point. The design of this type of transmission requires great care because the reflected wave can cause interference. Because the signal is being reflected off the earth's surface (or other surface), the path is longer. Therefore, the reflected signal can arrive later than the direct LOS signal. This delay, or out-of-phase signal, can distort the transmission. Figure 1-9 illustrates VHF transmission using a LOS signal. The reflected wave is also shown. In order to achieve LOS transmission, antenna height is critical; the greater the distance separating the transmitter and receiver, the higher the antenna.

At the *ultra high frequency* (UHF) band, the use of microwave signals is more prominent. In the microwave systems of today, high-range frequencies are used for point-to-point communications. Several channels of communication can be multiplexed together and transmitted across the carrier. Telcos, PTTs, and private organizations use these microwave systems

Figure 1-9
VHF transmission

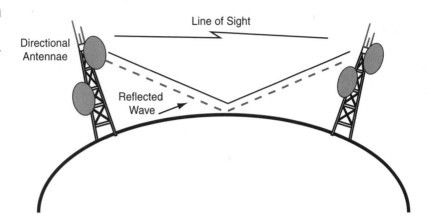

Line of Sight

Directional
Antennae

Reflected
Wave

extensively to carry telephone calls. Two sets of frequencies are required in microwave systems: a transmit frequency and a receive frequency. The lower bands of the frequency spectrum (LF and HF) are used as a one-way alternating transmission on a single frequency requiring one party to listen while the other speaks. A push-to-talk transmission is used. If both parties try to send at the same time, a jamming effect will render the transmission useless. Very specific transmission protocols are required so that the radio can be used effectively.

In a telephone call, the transmission of voice calls across a radio signal is different. The dynamics of a voice telephone call require simultaneous transmission in both directions because the protocols or rules of telephony are far less stringent. To avoid the jamming effect, two separate frequencies are required. Figure 1-10 shows a microwave transmission using two separate systems (a transmitter and a receiver). Each of the systems has a separate frequency. Frequency one is used to transmit from west to east, whereas frequency two is used to transmit from east to west.

Newer microwave systems combine these transmit and receive functions into a single device called a transceiver, which mixes the input-output onto a single radio system for two-way communication. Figure 1-11 illustrates the newer transceiver system.

A two-way radiotelephone system (such as *Advanced Mobile Phone Service* [AMPS] or *Improved Mobile Telephone Service* [IMTS]) and a cellular telephone system operate along the same lines as the microwave radio system. One difference is that radiotelephony and cellular systems use the lower frequency spectrum, so that the distance limitations and transmit output power needed apply. However, these systems now have dual

Figure 1-10
Two frequencies used on separate transmitter and receivers

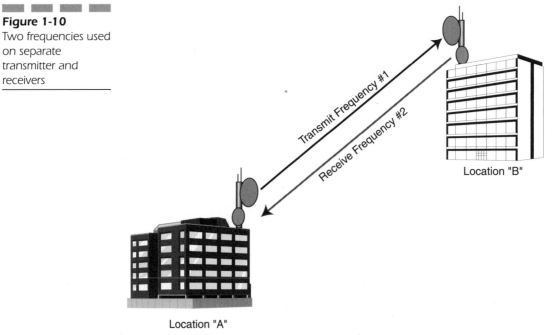

Transmit Frequency #1

Receive Frequency #2

Location "B"

Location "A"

Figure 1-11
Newer microwave systems combine the transceiver function.

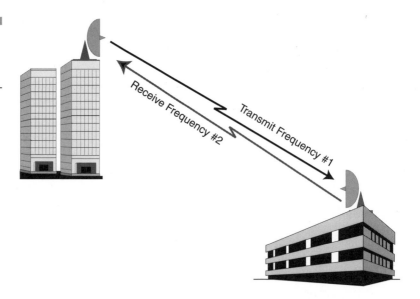

Receive Frequency #2

Transmit Frequency #1

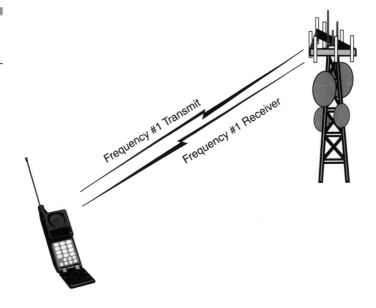

Figure 1-12
Two-way radio
systems

frequencies to connect to a radiotelephone transmission system. Figure 1-12 shows this general concept for a two-way radiotelephone system. Cellular phones work similarly, but use different terminology. The base station or cell site (depending on the system used) transmits at much higher wattage output than the mobile set (or cellular phone). Output at a base station or cell site can be in the 15- to 30-watt range, whereas the mobile set will output at 5 watts for older systems, 3 watts for cellular, and .3 to .6 watts for portable cellular sets.

Microwave Repeater Systems

As radio-based systems evolved, it was discovered that a microwave transmission could cover greater distance with repeaters. A repeater receives the radio signal on one antenna, converts the signal back into its electrical properties, and then retransmits the signal out across a new transmitter. In some cases, the repeater changes the signal from one frequency to another. Line-of-sight transmission is used, and the height of the antenna is important to maintain this capability. However, limitations from power, path, and other sources still restrict the distances covered. The distance limitations vary for microwave transmissions in various frequency ranges. Table 1-1 shows the normal distances for several frequency ranges before repeaters are needed.

The distances given in Table 1-1 are representative only. Clearly, each system will vary depending on external factors. For example, a 2- to 6-GHz radio installed atop Mt. Killington in Vermont could successfully achieve a 50 to 70 mile distance between systems without repeaters. The antenna's height in relation to the earth's curvature enables greater transmission distance and minimizes interference from other sources. In addition, the power output can be greater (although not required) at this height because nothing else is around.

Figure 1-13 shows a microwave repeater system at a typical distance of 30 miles. The system is for illustrative purposes only because other considerations would affect transmission. In the telephone systems across the world, this type of transmission has been used extensively. It is still widely used in networks across great distances.

Table 1-1

Distances

Band	Distance
2–6 GHz	30 miles
10–12 GHz	20 miles
18 GHz	7 miles
23 GHz	5 miles
38 GHz	1–2 miles

Figure 1-13

Microwave repeater systems

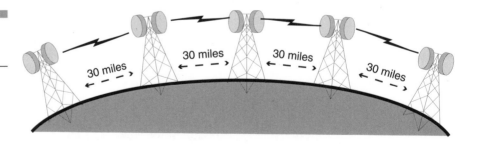

Satellite Radio Communications

As telephone systems continued to evolve and the need to transmit information over greater distances grew, a newer radio-based system emerged. In 1960, microwave radio signals were transmitted up into the atmosphere to a repeater floating in space. Called satellite communications, the system was originally designed to bounce radio waves off an artificial object that was orbiting the earth. Earlier attempts to bounce radio waves off the moon were not as successful as had been hoped. The system worked, but the returning signal was so weak that it couldn't be used. A similar attempt to bounce radio waves off an inflated weather balloon met with equally unrewarding results. Therefore, an active rather than a passive system became the next logical step.

An orbiting satellite offers several distinct advantages. A stronger signal can be obtained and over very long distances, the transmission signal requires only a single repeater (the satellite). The satellite can be located in a polar, inclined, or equatorial orbit. The orbit can be either circular (equidistant) or elliptical (non-concurrent) at different heights above the earth's surface. The early satellites were launched into elliptical orbits at lower heights above the earth than satellites of today. The orbit around the earth took from one to two hours, depending on the height and path. Therefore, the earth station ratio equipment had to track the satellite and could transmit only for limited periods when the satellite was visible. The system proved impractical for commercial use because several satellites were needed to provide constant communication and the moving antenna equipment required the constant re-aiming of dishes. Therefore, a circular orbit around the equator (equatorial orbit) at a height of 22,300 miles was selected. A satellite at this height takes 24 hours to orbit the earth, resulting in what looks like a stationary object. In fact, this is called a geostationary or geo-synchronous orbit.

Furthermore, at this height, a wide footprint (area of coverage) of the radio beam (at 170-degree dispersion) back on the earth's surface produces a very large area of coverage—approximately one-third of the earth's surface—from a single satellite. Thus, only three satellites are needed to provide almost total coverage.

Once the position of the satellite and the coverage areas are determined, the rest is straightforward. A microwave transceiver communicates with the satellite and its associated footprint. A single satellite can embrace North and South America in its radio beam path.

A terrestrial link (microwave, copper, coaxial, or fiber) is run from the customer's location to the nearest uplink for transmission and reception of satellite communications. Figure 1-14 shows a U.S. connection linked through various means. Internationally, the arrangement will differ, depending on the degree of access allowed in a particular country and the rules of its PTT organizations.

Satellite transmission has been used extensively for long-haul communications and for transoceanic international communications. However, the costs associated with satellite launch, electronic components, earth station equipment, and orbit life have all been somewhat detrimental in the overall picture. Although other technologies (such as microwave and fiber optics) offer declining scales of cost, satellite costs have remained high. For

Figure 1-14
Satellite links

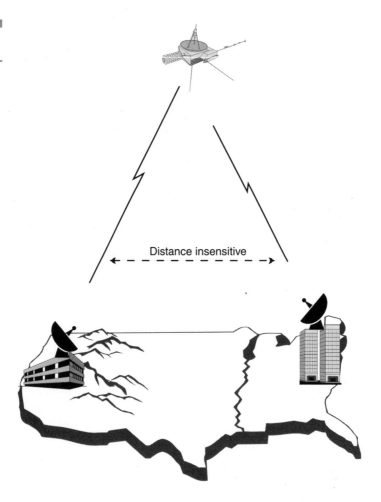

Distance insensitive

this reason, many industry users opted to use this transmission system selectively. Domestic use of these systems was expected to fade quickly into the past.

A large amount of bandwidth is available with satellite radio systems because the frequency spectrum can be allocated on a fixed-access or demand-access basis. Suffice it to say the capacity is available if the user wishes to experiment or use the system full time. Because the signal must be transmitted up 22,300 miles to the satellite, the rental cost of a channel (or other portion of the bandwidth) is insensitive to distance. For now, it is safe to assume that the benefits outweigh the transmission disadvantages; therefore, the satellite will be around for a long time to come.

Standards Groups

In the North American community, many organizations coordinate the standards and the allocation of frequencies. The standards committees define the interoperability of equipment produced by different manufacturers. These committees consist of members from government agencies, manufacturers, and end-users. Figure 1-15 is a representation of some of the standards committees influencing the use of radio frequencies.

ANSI

The first standards organization is the *American National Standards Institute* (ANSI), a non-government agency that sets the standards for the voluntary usage in U.S. and Canadian markets. ANSI establishes committees to define standards for products and services offered by the carriers and manufacturers based on perceived user need. Furthermore, ANSI establishes the rules on how the committees themselves operate.

TIA

The *Telecommunications Industry Association* (TIA) is a trade organization made up of the manufacturers and carriers. TIA provides materials, products, and system distribution services to the U.S. and countries around the globe. TIA represents the telecommunications industry along with the *Electronic Industries Association* (EIA), the *Cellular Telecommunications*

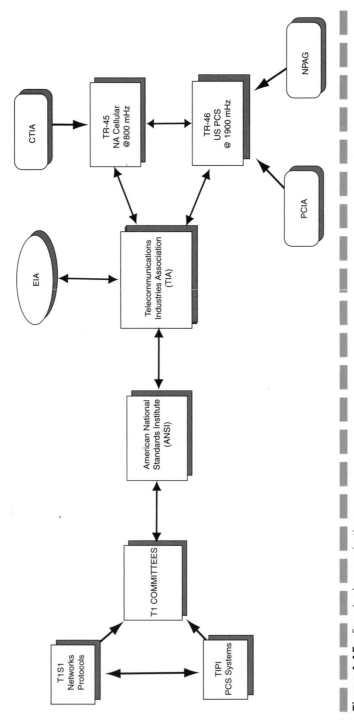

Figure 1-15 Standards organizations

Industry Association (CTIA), and the *Personal Communications Industry Association* (PCIA). TIA committees are broken into the following categories as shown in Table 1-2.

The next committee is the TR46 subcommittee that deals with issues associated with mobile and personal communications in the 1,800-MHz frequency spectrum. These include the following as shown in Table 1-3.

ECSA

The *Exchange Carriers Standards Association* (ECSA) subcommittees include the T1P1 and T1S1 standards. They are responsible for the standardization of services for networking switching systems together. The responsibilities of these committees are shown in Table 1-4.

Table 1-2

The TIA45 Committees Address Issues in the 800-MHz Cellular Spectrum

TIA Committee	Function
TR45.1	EIA-553 Analog Cellular
TR45.2	IS-41 Cellular Inter Systems Operations
TR45.3	IS-54/IS-136 Digital Cellular using TDMA
TR45.4	Micro cellular/ Personal Communications Services
TR45.5	IS-95 Wideband Spread Spectrum Digital Technologies (CDMA)

Table 1-3

The TIA 46 Committees Responsibilities

TIA Committee	Function
TR46.2	Network interfaces used
TR46.5	PCS 1900
TR46.6	Composite TDMA/CDMA standards

Table 1-4

Committees on ECSA That Have Roles in Wireless Communications

Committee	Function
T1P1	Lead role in PCS 1,900, Composite CDMA/TDMA, Wideband CDMA (W-CDMA), and PACS
T1S1	Works on ISDN and SS7 standards

Spectrum Usage

The demand for radio frequencies is high in most parts of the world. This means that controls must be placed on the use of the limited frequency spectrum. To coordinate the use of RF around the world, all radio services are licensed. The controls maximize the use of the frequency bands and minimize the occurrence of interference by limiting the access to authorized users who are held accountable for any interference that they may cause.

Regulatory Control

The *World Administrative Radio Conference* (WARC) is a global administration that meets every four years to review the latest radio technological advances, identify the need for new services, and coordinate the international operation and use of radio frequencies.

Local governments control the use of the radio frequencies in their respective countries. These governmental agencies assign available spectrum to specific applications such as for broadcast radio and TV, emergency services, and amateur radio.

The regulating bodies in each country are responsible for licensing radio services to ensure that they comply with national and international standards. Some of the regulatory bodies include

- **United States** *Federal Communications Commission* (FCC)
- **Canada** *Canadian Radio and Television Commission* (CRTC)
- **Europe** *Conference of European Posts and Telecommunications* (CEPT)

The spectrum being used in the analog radio environment is shown in Figure 1-16, where the various parts of the frequency spectrum were allocated for the U.S.-based operators. Note that this covers the range of the IMTS frequencies initially assigned, through the AMPS, and some of the Cellular RF spectrum covering *Total Access Communications Services* (TACS). Each of these components is discussed.

Cellular Service

Early mobile telephone services used the 40-MHz, 150-MHz, and 450-MHz frequency bands. Despite being mobile communications services, the

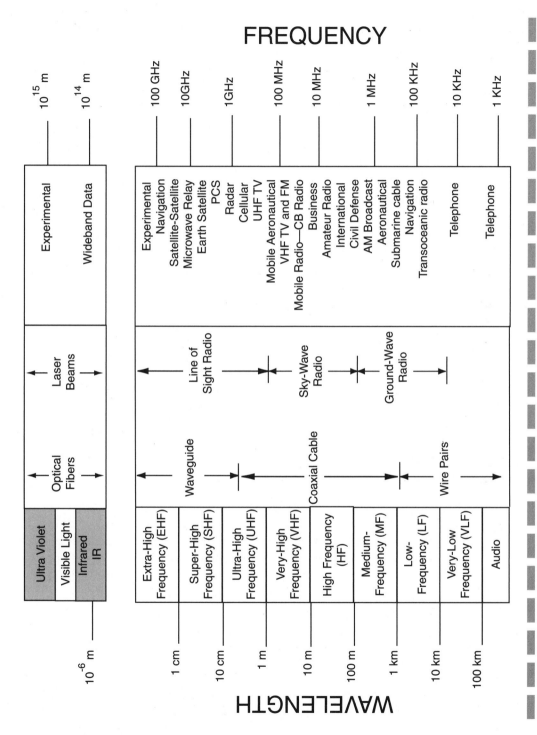

Figure 1-16 Analog radio spectrum for the U.S.

subscribers only got service in very limited areas. Cellular service, introduced in 1984 commercially, added frequency agility in mobile communications that allowed calling patterns across greater coverage areas.

Analog Cellular (AMPS and TACS)

Advanced Mobile Phone Services (AMPS) is a standard that was developed jointly by AT&T and Motorola. This service opened new services across the U.S. in the 800 MHz frequency band and took advantage of spectrum that was available after the demise of the UHF television industry.

In other parts of the world, cellular services were introduced in the 900 MHz frequency band. This band was already used in the U.S. Therefore, U.S. manufacturers had to find a different band; thus, the 800 MHz was used for the U.S.

In the United Kingdom, British Telecom developed a service called Total Access Control System (TACS) that was adopted by many countries in Europe and Asia. Many other standards and services were developed around the world, but these were the more common ones.

AMPS and TACS introduced high-quality voice communications, but limited roaming capabilities and very limited data operation. Security was almost an oversight. As the analog cellular movement caught on, subscribers experienced delays, disconnects, problems with roaming, and expensive billing arrangements. These subscribers began to pressure the carriers and manufacturers to improve the service. However, because of the rapid growth and acceptance of cellular communications, not enough spectrum was available to satisfy the demand. Diverse analog standards implemented throughout Europe made it possible to use the same phone when the subscriber traveled across the continent. However, in the U.S., this was not to happen because of inconsistencies in implementations.

Motorola developed a temporary solution to the spectrum problem called *narrowband AMPS* (NAMPS) until the migration and standardization to digital could occur.

Digital Cellular

To enhance the services of the AMPS network (increase capacity, address security, and roaming issues), the digital evolution of the cellular networks began. The new standards enabled multiple-caller support on the same number of frequencies, increased security (decreased fraud and cloning), and paved the way for data communications on the wireless cellular networks.

On the other hand, in the European market, the GSM standard operating in the 900-MHz frequency bands evolved as a replacement for the older analog systems. In North America, the *Digital AMPS* (DAMPS) was introduced to operate side by side with the analog systems (AMPS) in the 800-MHz frequency band.

Personal Communications Services (PCS)

The latest service offerings included the introduction of personalized communications capabilities, enabling the customization (personalization) of services for the end user. PCS operates in the 1,900-MHz and 1,800-MHz frequency bands using all digital services. PCS includes the use of voice, data, and messaging services combined with a potpourri of other features and functions. PCS is a worldwide standard; in North America, FCC introduced Personal Communications Network (PCN).

Universal Mobile Telephone Systems

Universal Mobile Telephone Systems (UMTS) promises to open more capacity and service offering on a worldwide basis as a standard operating in the 3,200-MHz frequency band. The UMTS systems are evolving and promoted by many of the manufacturers and operators alike.

A timeline of the implementation of these standards is shown in Figure 1-17.

Radio Channels

A critical component of any radio-based system is the capability to channelize the frequencies into smaller components. The channelization specifies how much of the spectrum each user is allowed to occupy during their selective use of the frequency. Included in this channelization is the spectrum allocated to the individual handset and the base station. All of the wireless networks work with a bi-directional communications flow, (duplex operation). Two-way communication is possible through this planning process.

Figure 1-17
Timeline of
implementation

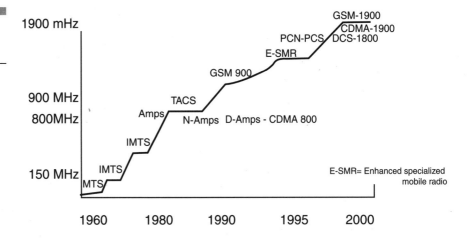

As communications take place, they will occur as follows:

- Transmission from the base station to the mobile set is called forward channel operation and uses the downlink.

- Transmission from the handset to the base station is called the reverse channel operation and uses an uplink.

Separating the up and downlink channels is called duplex separation, which happens to be the same for each of the channels. Table 1-5 is a sample of the frequency spacing for duplex separation using the various standards discussed.

Moreover, when the channels are operational, the frequency spacing allowed per channel (how much bandwidth is allowed per channel) is shown in Table 1-6.

Table 1-5

Spacing for Duplex
Operation by
System

System	Separation
Analog Cellular (AMPS/DAMPS/CDMA 800 MHz)	45 MHz
GSM (900 MHz)	45 MHz
DCS 1,800	95 MHz
PCS 1,900	80 MHz

Table 1-6

Channel Capacity
for Calls

System	Channel Capacity
Analog Cellular (AMPS/DAMPS 800 MHz)	30 kHz
CDMA (CDMA 800/ CDMA 1,900 MHz)	1.25 MHz
GSM (GSM 900/ DCS 1,800/ PCS 1,900 MHz)	200 kHz

Figure 1-18 is a sample of the frequency allocation for the 800-MHz AMPS spectrum showing the various channels and the separation for up and downlink capabilities. This is a graphic representation of the overall operation of the networks using analog cellular operation. Figure 1-19 is the spectrum using the GSM operation, and Figure 1-20 is the frequency allocation for the digital PCS 1,900-MHz frequency band.

Figure 1-18
AMPS spectrum
allocation

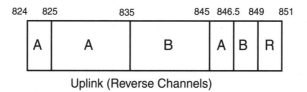

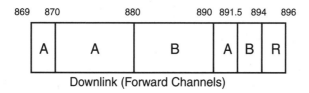

Figure 1-19
GSM 900 frequency
bands

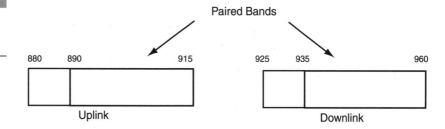

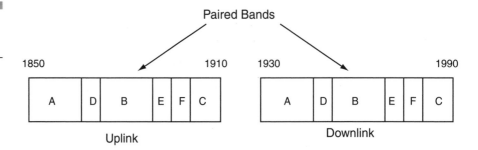

Figure 1-20
PCS 1,900 frequency
allocation

Multiplexing Arrangements

No matter how much bandwidth is allocated per standard, it is never enough! Therefore, each of the digital standards defines a way of channelizing the frequency band, but also to multiplex many simultaneous users on the spectrum. All of the digital standards define a multiplexing method (which does vary) to share the frequencies. The idea of the multiplexing standards evolved in three main thrusts. The standards shown in Table 1-7 evolved to handle the multiplexing for the different types of systems.

Looking at the different ways of multiplexing and providing full duplex operation requires that the systems must operate within the same guidelines based on the selection made. The various multiplexing techniques are discussed so that a basic understanding is attained of how they are used in a wireless environment.

Frequency Division Multiple Access (FDMA)

Primarily, the use of FDMA is an analog technique used by the original cellular providers to isolate each channel and conversation based on a single frequency. As a cellular caller is ready to place or receive a call, a pair of

Table 1-7

Types of
Multiplexing
Used

Concept	Techniques Used
Frequency Division Multiplexing (FDM)	*Frequency Division Multiple Access* (FDMA)
Time Division Multiplexing (TDM)	*Time Division Multiple Access* (TDMA)
Code Division Multiplexing (CDM)	*Code Division Multiple Access* (CDMA)

frequencies is scanned and selected for the duration of the call. In frequency division, the spectrum is sub-divided into the individual 30-kHz channels. The channel is not fixed to a specific user; the multiple access means that users compete for the channel on a first-come, first-served basis. Once the channel is released (after the call is ended), it becomes available to the next user who may wish to make or receive a call. Each channel is assigned to a frequency pairing. In Figure 1-21, the frequency pairs are shown as availability permits. The caller is granted the entire channel for the duration of the call.

Time Division Multiple Access (TDMA)

TDMA is different in that transmissions are formatted into a frame. Frames are divided into specific time slots. Each call/message is assigned to a specific time slot and is only allowed to occupy the channel at that time only. Because several time slots are made available per frequency, several time slots are created and several users can occupy the same frequency but different time slot simultaneously. The number of callers that can be serviced at the same time on the same frequency is multiplied, as shown in Figure 1-22.

DAMPS and GSM both use TDMA, but different formats of time slotting have been adopted. Each communications mode is supported on an assigned pair of frequencies to support time division duplexing (full-duplex operation).

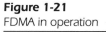

Figure 1-21
FDMA in operation

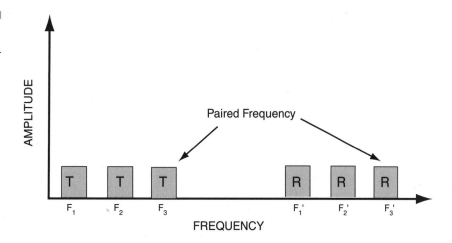

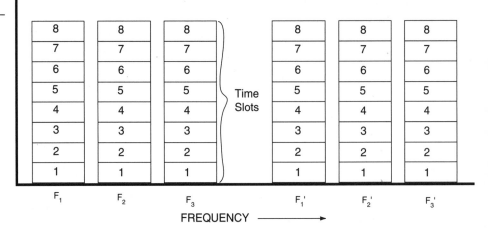

IS-54 DAMPS standards use TDMA to multiplex conversations on the voice channels only, and use a separate setup channel that is the same as AMPS. The latest and greatest form of digital AMPS is IS-136 that incorporates time slots on the setup channels and the voice channels.

GSM uses TDMA to multiplex both voice and setup time slots on the same channels. A single time slot is used for call setup on small systems. However, for larger systems, a total of eight time slots can be used for this purpose. The contrast of the DAMPS and GSM systems is shown in Figure 1-23.

Code Division Multiple Access (CDMA)

CDMA uses a direct sequence spread spectrum technology developed by Qualcomm and Inter Digital Corporation to modulate voice information. The voice is encoded at 8 Kbps or 13 Kbps (depending on the technique used) and spread across a very large channel capacity. This spreading signal will cause the transmitted signal to increase to an output of approximately 1.28 Mbps. Figure 1-24 is a representation of the use of CDMA.

Using noise with known characteristics, the transmitter modulates the radio signal. Pseudo-random noise used is the result of applying a Walsh code (there are 64 predefined codes) to modulate the voice. The 64-bit code ensures security, prevents duplication of the signals, and makes sure that

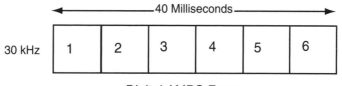

Figure 1-23
TDMA uses different modes.

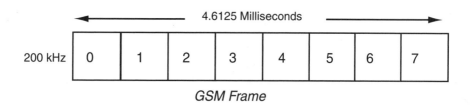

Figure 1-24
CDMA in action

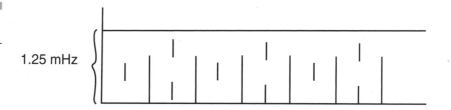

only the desired recipient can demodulate the information from the channel.

Each CDMA channel uses 1.25 MHz of RF in both the uplink and downlink channel direction. Despite the large bandwidth, CDMA can carry 10 times more calls than the old AMPS networks. The bandwidth is not an issue because the spreading signal is modulated so quickly that little to no interference occurs from adjacent users on the same frequency band. Theoretically, an unlimited number of users can occupy the same channel simultaneously because of the spreading and the Walsh coding. However, as more users log on, the noise level increases and the quality of the calls decreases proportionately. The more noise on the system, the fewer users allowed. The cell sites will begin to hand off the calls to adjacent cells or drop the calls.

Modulation Techniques Used

Many variations of modulation are used to prepare the information for the airwaves. The more common forms of modulation are as follows:

- **Amplitude Modulation** AM or ASK
- **Frequency Modulation** FM or FSK
- **Phase Modulation** PM or PSK
- **Quadrature and Amplitude Modulation** QAM or QPSK

The more common techniques used today include the FM and the QAM techniques for the providers. As we saw in the overall operation of the RF spectrum, several frequency bands are used for the transmission of radio signals. Looking at the mapping of the RF spectrum shown in Figure 1-25, we see the differences between the ways the radio frequencies are allocated and how the wavelength is used.

Radio signals will disperse over distance to fill in the space into which they are radiated. Signal strength will degrade over distance at a pre-

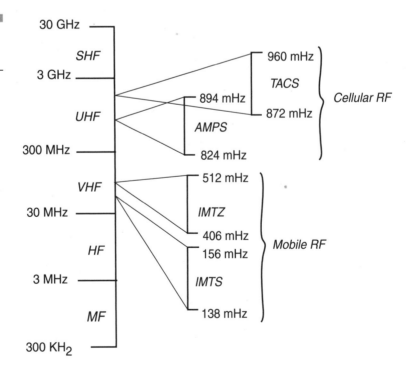

Figure 1-25
Frequency bands allocation

dictable rate. The further the transmitter and receiver are apart from each other, the greater the risk of signal degradation. This all depends on other conditions like the terrain and man-made structures that will get in the way, as we saw earlier in this chapter. Using the various parts of the spectrum therefore depends on the availability of frequencies in a specific area and the equipment that modulates the signal into the air.

Modulation is the technique used in radio communications to change the message to be sent into a suitable form and format. The format must accommodate the natural characteristics of the medium (in this case the airwaves). The information signal containing the message is altered (changed) to carry the message. This change agent is the modulator that will change the shape or the format of the electromagnetic wave to carry the signal appropriately for the transmission equipment. Modulation is a reversible function; the receiver demodulates the carrier wave to extract the information from the electromagnetic wave. Thus, we use a modulator and a demodulator to handle the transfer of information.

Modulated Signal Envelope

When we use a radio-based system to carry information (voice, data video, and so on), the radio system uses a base carrier. This is an unmodulated carrier where a constant carrier tone is sent between the point-to-point radio systems. From there, we add our information, which is then modulated (changes the base carrier frequency) with the information and the carrier frequency creating a modulated envelope with the voice (or whatever we are transmitting) on the radio wave. This envelope is transmitted to the receiving station, where the base carrier frequency is extracted and the information is what remains, as shown in Figure 1-26.

Figure 1-26
Modulated signal
envelope

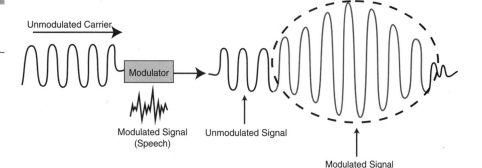

This sounds complex, but it is a tried and proven transmission capability. The real issue is that the information can be modulated on and off from the basic signal. This enables us to carry the voice channels. Voice channels are multiplexed onto the radio signal from an input multiplexer such as a D4 channel bank or an ADPCM channel bank or mux.

Amplitude Modulation

When the signal is being modulated onto the radio frequency, this service can be provided in many ways. The initial analog systems in radio frequency transmission used *amplitude modulation* (AM). In AM, the information is modulated onto the signal by changing the amplitude of the signal. The frequency is kept constant. The changes shown in Figure 1-27 are amplitude modulated.

Remember the two constantly changing components of the electromagnetic analogous transmission included both amplitude and frequency changes that occur in a time domain. If we keep the frequency constant, then we can provide the changes (modulation) to the amplitude of the base carrier. The receiver then is looking for amplitude shifts, and we call this *amplitude shift keying* (ASK). The problem with an AM system is the propensity to be degraded by amplitude changes. Noise on a radio wave comes in the form of amplitudes or increases in noise levels. As a result, the AM systems are more prone to error and noise interference.

Think about the AM radio in your car. You note that the quality is not as good as the FM radio stations. As you drive further from the radio station,

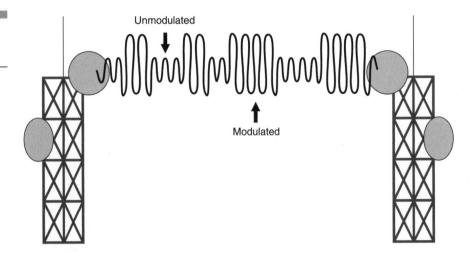

Figure 1-27
Amplitude
modulation

the quality drops significantly. The electrical interference increases in the reception of the signal causing static and noise on the radio wave (the songs). So what do you do? Crank up the volume! This is so you can amplify the received signal. Unfortunately, it works just the opposite to the expected result; the more you amplify the signal, the more you amplify the noise. This makes the received signal intolerable, so you finally do what you should have done in the first place—change the channel and look for better reception.

Frequency Modulation

The second choice is to modulate the frequency and keep the amplitude constant. The carrier wave's frequency is changed to carry the message. Because it does not rely on the envelope of the carrier to transmit information, it is relatively immune to the noise that impairs the AM. *Frequency modulation* (FM) is used in analog cellular radio systems, commercial radio broadcast, and most other modern two-way radio systems.

As the amplitude of the information changes, the frequency of the carrier changes at the same rate. Positive polarities of the information signal cause the frequency to increase, whereas negative polarities cause the frequency to decrease. The range of frequency change is called carrier deviation. Deviation cannot be allowed to drive the carrier frequency beyond the bandwidth of the channel or else it will interfere with neighboring channels. Frequency modulation is shown in Figure 1-28.

Figure 1-28
Frequency
modulation at work

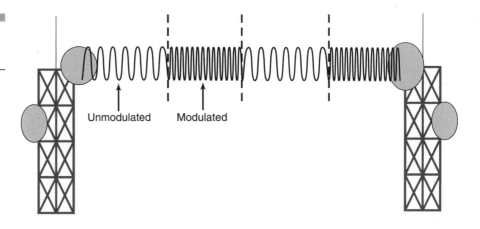

Unmodulated   Modulated

Digital Modulation

The main advantage of digital modulation techniques is improved resistance to noise, or robustness. In analog systems, every component of the network adds some degradation to the signal. Digital systems offer noiseless, error-free voice and data services by detecting distortions or losses and correcting for them before they are delivered to the end user terminal.

Digital modulation is used to impress the information contained in a digital signal into an analog carrier wave. A time relationship exists between the square pulses of the digital information and the various digitally modulated carrier waves.

Some examples of digital modulation techniques used in a radio system are shown in Figure 1-29 and are as follows:

- **Frequency Shift Keying (FSK)**   Where the frequency of the intelligence or the carrier is changed from one value to another, to indicate data states. This method is used in the AMPS control channel to convert binary data into analog tones compatible with the analog carrier.

- **Phase Shift Keying (PSK)**   Where the phase of the carrier is switched between states. A two-state modulator transmits single-bit values only. Modern systems use multi-state modulators to transmit multiple-bit values (chips) with each phase change.

Figure 1-29
Digital modulation
techniques

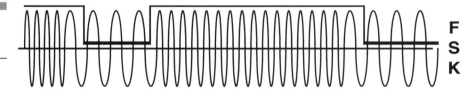

F
S
K

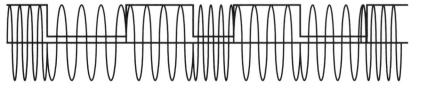

P
S
K

Light-Based Systems

Another wireless transmission system that is both reliable and inexpensive has proved to be less popular than radio-based systems. Infrared light has long been used as a transmission medium for short-haul local communications. Although the benefits of radio are many, the quality impairments, limited-frequency spectrum, and rain attenuation problems all must be considered equally as disadvantages. The primary advantages of infrared systems are that they are easily transportable, typically can be set up and operating within an hour, and do not require expensive licenses or right-of-way permits. For all these advantages, the main disadvantages are the limited distance and limited bandwidth. In addition, infrared transmissions are subject to even more fading and absorption than microwave systems.

A pinpoint-beam generator in the infrared (invisible) light spectrum operates at a frequency range of 10^{12} to 10^{15} and a wavelength of 10^{-4} to 10^{-10} to produce high-speed communications. It should also be understood that in this context, infrared is carried in open airwaves. If a different medium, such as if very pure glass (fiber) were used, the spectrum of the light would be in the infrared range. The fiber achieves the high-speed, reliable transmission because the impairments of the air are overcome on the glass.

Most consumers are familiar with wireless infrared transmission although they may not realize it. The remote controllers for TV sets, radios, lights, and other appliances use infrared signaling. Clearly, these low-powered controllers are very limited applications, but they have all the properties of infrared transmission. They can be used only for short distances and are directional (point-to-point), invisible light transmitters.

In addition, these systems are one-way devices; separate transmitters are needed for two-way communications. Most providers recommend distances of less than 1 mile to maintain the reliability and integrity of infrared transmission.

Infrared systems also have a place in the voice and data communications arena. Overall, the signals can be used effectively to handle a variety of inputs across a spectrum to deliver 4, 10, 16, 25, 100, and 155 Mbps. Newer wireless infrared systems use the higher end 100- to 155-Mbps capacities to link point-to-point needs, and 622 Mbps are now on the market.

Broader band communications now offer a free-space (wireless) transmission at optical speeds similar to an OC-48 (2.4 Gbps). However, the future promises speeds of up to 10 Gbps within 3 years. In general, we can expect to see more light-based systems creeping into carrier and end-user networks.

Radio Technologies and Systems

Wireless Methodologies

In the introduction of Chapter 1, the basic concepts of the wireless history were undertaken. The history provided an overview of the evolution of wireless techniques. This chapter discusses some of the basic physics associated with wireless transmission techniques and describes how radio technology is implemented in various applications. This description includes a more in-depth look at some of the networking technologies mentioned in Chapter 1.

We will look at the various differences in the mode of operation for ground-based systems and air-based systems. We will also see the differences in coverage using the various wavelength operations of a long wave, short wave, and microwave. Finally, we will address the areas of satellite communications systems dealing with the varied use of the differing orbits that they occupy. They are referred to as the *xEO*, where *x* defines the height of the orbit (LEO is *low-earth orbit*, MEO is the *mid-earth orbit*, and GEO is a *geosynchronous orbit*.)

Radio Frequency Characteristics

The characteristics of radio waves in any frequency band determine how useful those frequencies are for the service required. The main characteristic of interest is how signals are changed or distorted by absorption and reflection and by the air and other physical media before reaching the receiver. Because a radio signal is a particular form of energy, it is useful to consider the different forms of energy and how they can be converted and transmitted.

General Aspects

In order to gain an appreciation for some of the properties of radio frequency propagation, it is important to define the underlying principles or the physics of electromagnetic waves. (Oh no! This sounds very technical!) Energy must be moved to create the transmission. Energy is defined as the result of multiplying power by time. It can take a number of different forms, such as

- Sound energy
- Electrical energy

■ Kinetic energy

■ Potential energy

■ Heat energy

■ Chemical energy

A fundamental principle of physics is that of the conservation of energy:

"Energy cannot be gained or lost; it can only be converted from one form to another."

When we talk about generating a particular form of energy, say electricity, what we are really doing is changing one form of energy (heat) to another form (electricity) in a power-generating station. When one form of energy is converted to another, the process may not be the most efficient form of changing the energy into the new form. When we convert chemicals into other forms of energy, a certain amount will be changed in a side effect (meaning that we do not get 100 percent transference). Some of the energy is not lost but changed into a different form. When we generate electricity at the power station, we are converting chemical energy from fossil fuels into electricity. However, a certain amount of the energy is displaced in the form of heat.

A special form of kinetic energy is heat. The temperature of an object is due to minute vibrations of its molecules. As an object cools, it will transmit its heat energy to its surroundings, air and other objects, by means of infrared radiation and conduction. Eventually, the object and its surroundings will all have the same temperature and no energy will be transferred. This is known as *entropy*, which is the tendency for all energy to have the same potential, and thus no useful work can be done. Sound can also be looked upon as a form of kinetic energy, but in the form of pressure waves or vibrations in a fluid or solid. Because sound requires a medium, it cannot traverse the vacuum of space. Thus, we need to use a medium to create the carrier of the energy we create. This can be in the airwaves through RF modulation, in the wires as electricity, or though some other combination of the various forms.

Electrical energy is transferred by one of two means:

■ **Conduction**   When an electric current flows through wire, energy is transferred by conduction.

■ **Radiation**   Electrical energy is also radiated by a radio transmitter.

An electric current will flow in a conductor such as a copper wire if a potential difference exists between the two ends. A potential difference can be considered as an excess of electrons at one end and a shortage of electrons at the other end. As the current flows, an electromagnetic field is generated. If the wire has resistance, some of the energy will be converted to heat, thus warming the wire and causing thermal noise.

Ohm first described the relationship between the potential difference (voltage), the electrical resistance of the wire, and the current flowing through the wire for direct current in his well-known law:

$$E = I \times R$$

Where:
E = Potential difference in volts
I = Current in amps
R = Resistance in ohms

The greater the resistance (which is proportional to the length) of the wire, the less current will flow through the wire for a given voltage level. This goes back to the old garden hose analogy that is used to describe bandwidth.

This law forms the basis for much of electrical theory and is valid for direct current (an electric current that does not vary in time). An electric current that varies in time is known as an *alternating current*, which is the normal method for distributing electricity for domestic and commercial use. When dealing with alternating current, the R (representing resistance) is replaced by a Z (representing impedance):

$$E = I \times Z$$

Where:
E = Potential difference in volts
I = Current in amps
Z = Impedance in ohms

Impedance is made up of two components:

■ R – The resistance of the wire

■ X – The active or frequency dependent part of the impedance, 90° phase shifted with R, where the following equation is used to describe the late relationship between them:

$$Z = R + iX$$

The *i* simply means 90° phase shifted.

This shows how impedance changes with the frequency of the signal passing through it. A high-frequency alternating current will generate a radio frequency signal as it passes through a conducting wire, creating the simplest form of a radio transmitter.

The different forms of electromagnetic radiation are defined by their frequencies. These include radio waves, infrared radiation (heat), visible light, ultraviolet light, X-rays, and gamma rays. All these different frequencies of electromagnetic radiation form the electromagnetic spectrum. In Chapter 1, we saw the various frequency ranges of the radio- and light-based systems. However, these are banded differently in Figure 2-1 where the frequencies are set aside for the various operations we plan to use in the higher-end microwave bands.

Consider radiant energy as an electromagnetic wave. Electromagnetic radiation can travel through free space and through various solids and fluids to varying degrees, depending on the frequency and the kind of solid or fluid. For example,

- Light can travel through air, water, and glass, but not other solid material.

- Radio frequency waves can travel through some solids, but not through metal.

- X-rays and gamma rays will not travel through metal.

Figure 2-1
The frequency ranges used by type and service

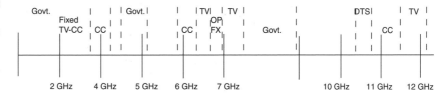

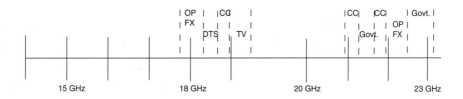

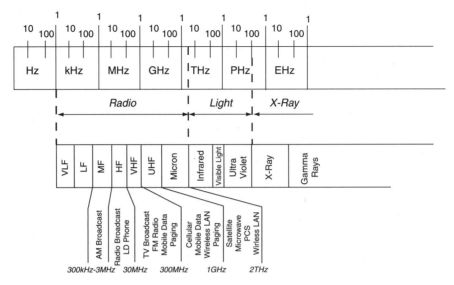

Figure 2-2
RF spectrum for radio
and light

Figure 2-2 is a more definitive look at the lower frequency bands up to the higher frequencies.

The higher frequency waves have more capability to penetrate solids than those with lower frequencies. Although radio frequency waves may be able to penetrate the material of a building, the construction of modern buildings may prevent radio transmissions from reaching the inside of an office block. Most modern buildings are constructed using steel beams for the main structural integrity. The external cladding (perhaps granite or marble) is fixed to the frame to enclose the space and provide an aesthetically pleasing appearance. Internal subdivisions for offices are constructed using steel or wooden frames to support partition walls. Radio waves are able to penetrate the cladding of the building, but the steel frame acts as a *Faraday Cage* to effectively screen the interior of the building to radio waves of some wavelengths. This effect was named after Michael Faraday who was the first to demonstrate and explain it. If the construction of the frame, or cage, is such that the spaces between the steel girders equate to or are smaller than the wavelength of a radio signal, then the signal is drastically attenuated. For use in buildings, radio frequencies must be carefully selected to ensure that the best compromise is made between the Faraday Cage effect and the material penetration capability of radio waves. The Faraday Cage effect is used in electronic devices to provide screening of unwanted radio frequency signals without the need to use solid metal enclosures. Figure 2-3 illustrates the Faraday Cage effect in a modern building.

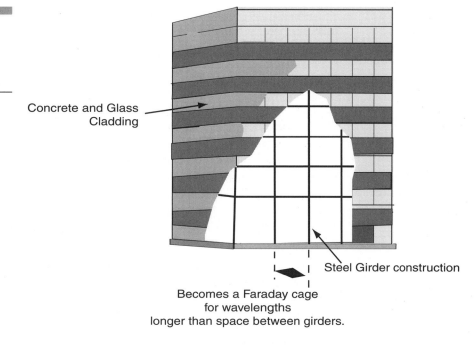

Figure 2-3
The Faraday Cage
effect in a modern
building
(Source: IBM)

Concrete and Glass
Cladding

Steel Girder construction

Becomes a Faraday cage
for wavelengths
longer than space between girders.

Contrary to popular belief, outer space is not empty. It is filled with *electromagnetic radiation* that crisscrosses the universe. This radiation comprises the spectrum of energy ranging from radio waves on one end to gamma rays on the other. It is called the *electromagnetic spectrum* because this radiation is associated with electric and magnetic fields that transfer energy as they travel through space. Because humans can see it, the most familiar part of the electromagnetic spectrum is visible light (red, orange, yellow, green, blue, indigo, and violet). Like expanding ripples in a pond after a pebble has been tossed in, electromagnetic radiation travels across space in the form of waves. These waves travel at the speed of light—just under 300,000 kilometers per second. Their wavelengths, the distance from wave crest to wave crest, vary from thousands of kilometers across (in the case of the longest radio waves) to smaller than the diameter of an atom, (in the case of the smallest X-rays and gamma rays).

Electromagnetic radiation has properties of both waves and particles. What we detect depends on the method we use.

■ The beautiful colors that appear in a soap film or in the dispersion of light from a diamond are best described as *waves*.

■ The light that strikes a solar cell to produce an electric current is best described as a *particle*.

When described as particles, individual packets of electromagnetic energy are called *photons*. The amount of energy a photon of light contains depends upon its wavelength. Electromagnetic radiation with long wavelengths contains little energy. Electromagnetic radiation with short wavelengths contains a great amount of energy.

Radio waves have the longest wavelengths, ranging from a few centimeters from crest to crest to thousands of kilometers. Microwaves range from a few centimeters to about 0.1 cm. Infrared radiation falls between 700 nanometers and 0.1 cm. (Nano means one billionth.)

In a vacuum, all electromagnetic radiation will travel at the same velocity: that is, 186,321 miles/s or 299,790 km/s. This is commonly termed *the speed of light*. The velocity in fluids and solids will vary according to the type of material and the frequency of the radiation. This can be easily demonstrated when white light is passed through a prism. White light is made up of a number of different frequencies corresponding to the different colors of the visible spectrum. The shorter wavelengths (higher frequencies) travel more slowly through the glass of the prism and are refracted more than the lower wavelengths. Thus, the violet light will be bent by the glass of the prism more than the red light; this will result in the white light being spread into a spectrum of all the colors, as shown in Figure 2-4. Some radio waves have similar properties to light and similar techniques may be used to control them. Electromagnetic radiation is normally considered to consist of a sine wave, which has the properties of wavelength, frequency, and amplitude. The visible part of the electromagnetic spectrum ranges from 400 to 700 nanometers and consists of the rainbow of colors (violet, indigo, blue, green, yellow, orange, and red, respectively) we observe when light is refracted through mediums of different refractive index, such as water or a

Figure 2-4
The prism effect
of light

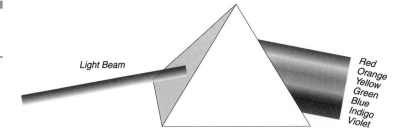

glass prism. For comparison, the thickness of a sheet of household plastic wrap could contain about 50 visible light waves arranged end to end. Below visible light is the slightly broader band of ultraviolet light that lies between 10 and 300 nanometers. X-rays follow ultraviolet light and diminish into the hundred-billionth of a meter range. Gamma rays fall in the trillionth of a meter range.

The relationship between frequency and wavelength is given by the following equation:

$$\lambda = (300 \times 10^6)/f$$

Where:
f = Frequency in Hz
λ = Wavelength in meters
(300×10^6 is the speed of light)

Electromagnetic radiation can be generated in various ways according to the frequency of the radiation required. Light and heat can be generated by simply raising the temperature of an object, whereas radio waves and X-rays require more sophisticated methods.

Objects that are raised to very high temperatures will radiate energy over a very wide range of the electromagnetic spectrum. For example, the sun radiates radio frequency, heat, visible light, ultraviolet light, X-rays, and gamma rays. However, it is not practical to use this method to generate and control anything other than heat or light.

Electrical energy is transmitted in the form of electrical impulses or waves, regardless of whether the energy is conveyed across wires, air, or water. The frequency is expressed in hertz (Hz), which represents impulses or cycles per second. The electrical energy, or signal, is changed by the medium that it passes through. It can be attenuated (absorbed) or reflected, resulting in a signal that is distorted in some way. Waves are changed in size or amplitude (attenuated), direction (reflected), or shape (distorted), depending on the frequency of the signal and the characteristics of the medium that they pass through. By choosing the correct medium, a signal can be changed or controlled. An electrical signal will be attenuated when it passes through a wire. High-frequency light signals travel through air, are reflected by mirrored surfaces, and are absorbed by most solid objects. For example, light signals pass through the atmosphere, but they are blocked by solid walls, unless made of glass or transparent material. Low-frequency signals do not propagate well by air, but can travel well through some solid objects depending on conductivity.

Frequencies below 900 MHz can generally propagate well through walls and other barriers. As radio frequencies increase and approach the frequency of light, they take more of the propagation characteristics of light. Signals between 900 MHz and 18 GHz, typically used by wireless LANs, are not as limited as light, but still do not pass through physical barriers as easily as typical radio broadcast band signals. Signals of 300 MHz or higher can be reflected, focused, and controlled like a beam of light. Parabolic-transmitting antennae use the properties of UHF and higher frequency signals to enable a relatively low-power signal to be focused directly toward its destination. Even closer to light signals, infrared signals have properties similar to light.

Choosing the most suitable frequency can determine the best propagation or transmission characteristics. The fact that only radio signals of certain frequencies are reflected by certain surfaces can be utilized to an advantage. For example, the capability of high-frequency microwave signals to penetrate the earth's atmosphere without being reflected is useful for satellite communications. Lower frequency signals (200 kHz to 30 MHz) are reflected back from the ionosphere (upper layer of the atmosphere), depending on time of day, season, and sunspot activity. This characteristic enables radio signals to be bounced off the ionosphere for long-distance communications beyond the horizon, such as using troposcatter radio systems.

When higher frequency carrier waves are used, more bandwidth is normally available to transmit information. By increasing the bandwidth of a communications channel, more data may be transmitted in a given period because the information is directly proportional to the bandwidth of the signal. The frequencies of most interest to wireless transmission range from near the 200 kHz mark, where long-wave radio transmissions are situated, up to infrared light in the terahertz range. Using higher frequencies has some drawbacks. The technology to build radio transmitters and receivers at higher frequencies is more complex. At higher frequencies, the wavelength of the radio signal approaches the physical length of the connections in the radio itself. Because a wire with lambda/four or multiples of this length is a good antenna, the actual connections within the radio itself must be kept short and become part of the circuit design because of problems with signal leakage. The path loss between transmitter and receiver is also a function of the wavelength:

$$\text{Path Loss in dB} = 20 \log 10 \ (l/4pR)$$

Where:
R = Range in meters
L = Wavelength in meters

Another property of electromagnetic radiation is that it can be polarized. Radio waves can be polarized in the same way, and selection of polarization of a transmitted signal may be achieved by the position of the transmitting elements in a horizontal or vertical attitude. This property can be used to reject unwanted or spurious signals, which may arrive at the receiving antenna with a different polarization to that of the wanted signal.

Microwave Communications

No one ever pays much attention to the microwave radio dishes mounted on towers, on the side of buildings, or any other place. This technology has been taken for granted over the years. However, this nondescript industry has quietly grown into a $3.4 billion global business annually. Four major suppliers provide one-half of all the radio-based systems globally. Microwave has also become a vital link in the overall backbone networks over the years. Now it has achieved new acclaim in the wireless revolution relaying thousands of telephone conversations from place to place, bypassing the local landlines.

Microwaves are between 1 mm and 30 cm long, and operate in a frequency range from 300 MHz to 300 GHz. Microwaves were first used in the 1930s when British scientists discovered the application in a new technology called radar.

In the 1950s, microwave radio was used extensively for long-distance telephone transmission. With the need to communicate over thousands of miles, the cost of stringing wires across the country was prohibitive. However, the equipment was both heavy and expensive. The radio equipment used vacuum tubes that were bulky as well as highly sensitive to heat. All that changed dramatically when integrated circuits and transistors were used in the equipment. Now the equipment is not only lightweight, but also far more economical and easy to operate. In 1950, the typical microwave radio used 2,100 watts to generate three groups of radio channels (each group consists of 12 channels), yielding 36-voice grade channel capacity. Each voice grade channel operated at the standard 4 kHz. Today, equipment from the many manufacturers (and Harris/Farinon specifically) requires only 22 watts of output to generate 2,016 voice channels. Although two orders of magnitude improvements have occurred in the quality of the voice transmission, the per-channel cost has plummeted from just over $1,000 to just under $37. This makes the transmission systems very attractive from a carrier's perspective. However, the use of private microwave radio has also

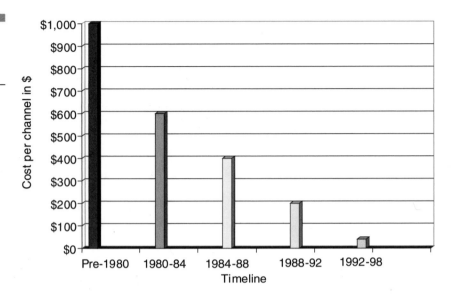

Figure 2-5
Comparison of cost per channel over the years

blossomed over the years because of the cost and performance improvements. This is shown in the graph reflected as Figure 2-5. This shows why the use of microwave has become so well accepted in the industry.

Today's microwave radios can be installed quickly and relocated easily. The major time delays are usually in getting through the regulatory process in a governmentally controlled environment. Several installations have taken over a year to be approved, only to have the radio system installed and running within a day or two. In many situations, microwave systems provide more reliable service than landlines, which are vulnerable to everything including flooding, rodent damage, backhoe cuts, and vandalism. Using a radio system, a developing country without a wired communications infrastructure can install a leading-edge telecommunications system within a matter of months. For these reasons, regions with rugged terrain or without any copper landline backbone in place find it easier to leap into the wireless age and provide the infrastructure at a fraction of the cost of installing wires.

The cellular and *personal communications service* (PCS) industries invested heavily in microwave radios to interconnect the components of their networks. This is shown in Figure 2-6 where the interconnection is used in the cellular world. In addition, a new use of microwave radio, called micro/millimeter wave radio, is bringing transmission directly into buildings through a new generation of tiny receiver dishes.

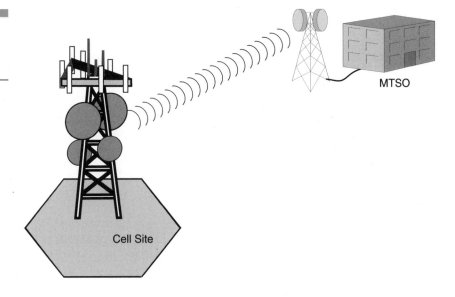

Figure 2-6
Cellular
interconnection of
microwave radio

Cell Site

MTSO

The new PCS industry is expected to choose microwave radio technology for the interconnection and backhaul transport on its expanding network. The PCS and the cellular suppliers do not want to pay the local telephone company for monthly T1 access lines from the cell sites to the mobile switching sites. Therefore, to eliminate the monthly recurring charges, they have installed microwave radio systems in the 18- to 23-GHz frequency range. Tens of thousands of new cell sites and PCS sites will be constructed over the next few years, further expanding the use of microwave radio systems in each of these sites. As third-generation handheld devices make their way into the industry, more wireless interconnectivity will be used.

Microwave also played a very crucial part of the PCS industry as the PCS systems use the 1.9- to 2.3-GHz frequency band. Fixed systems operators such as police, fire, electric utilities, and some municipal organization occupied these frequencies. To accommodate the move of these users from the 2-GHz frequency band, microwave was used to relocate the users to a new band, as mandated by the FCC. One study indicated that the PCS industry would spend over $3 billion in microwave equipment and services by 2005.

Another large demand for microwave emerged in the *Competitive Access Providers* (CAPs) market. CAPs offer long-distance access to customers at lower prices than the local telephone companies and the newer competitors. The CAPs normally install their own fiber-optic wires. However, they recognize the benefit of expanding coverage to consumer building entrances using a wireless high-speed connection. The CAPs are supplementing their fiber-based networks with Wireless CAPs (WCAPs). WCAPs use microwave transmission to deliver the Telecom service without the need for costly wire-based infrastructure. See Figure 2-7 for a graphic representation of this concept.

The newer micro/millimeter-wave radios, which are smaller and usually less expensive than other microwaves, are also popular with these CAPs and PCS suppliers. They are used in urban areas to extend the fiber networks. These radio units use the high frequency (or millimeter) bandwidth that hadn't been used. Now they are seen as a solution to increasing congestion in the lower frequency bands. An advantage of these systems is the small-size antennas that can be hidden on rooftops without interfering with zoning ordinances or creating aesthetic controversy.

Microwave is heavily used in radio and television systems. Satellite TV relies on microwave repeaters on the satellite to retransmit TV signals to a receiving station. Microwave communication via satellite provides a more reliable signal than longer, land-based radio waves. It also improves the reception of the picture.

Some TV stations have been using microwave to facilitate wireless communications from field cameras since 1992. What we continually hear about the *Action Cams* is a portable microwave system connected to a camera for

Figure 2-7
Wireless
interconnection with
fiber and CAPs

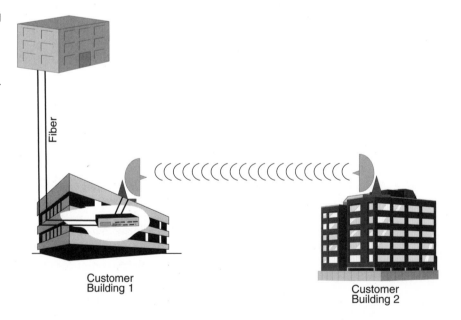

Fiber

Customer
Building 1

Customer
Building 2

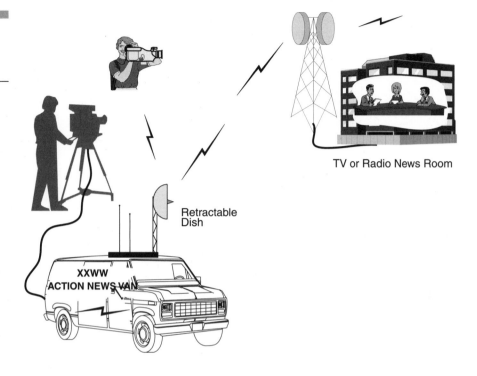

Figure 2-8
Action camera and
microwave systems
working together

TV or Radio News Room

Retractable
Dish

XXWW
ACTION NEWS VAN

real-time broadcast. Instead of being constrained to a fixed location, a news van can be driven and hooked up instantly, as shown in Figure 2-8. The systems hook up with a field camera with microwave units the size of a deck of cards. These can go anywhere and can operate from locations up to 2 miles from the van. Action and news is transmitted back to the van where it is relayed via microwave to the TV station.

What About Bandwidth?

Bandwidth is always a touchy subject. It can become a never satisfied drain on the corporate bottom line if due diligence is not practiced. A direct relationship exists between cost and total bandwidth. The more bandwidth needed, the greater the cost. Everyone would like as much bandwidth as possible and at the same time, have it be affordable. Many people make the mistake of buying more bandwidth than needed, anticipating future growth. In this industry, prices keep falling as competition increases. If an organization needs an OC3 (155 Mbps) today, then laying fiber is probably the most affordable solution.

Conversely, if 10 Mbps Ethernet is the current rate of transmission, this demand can be met immediately. Additional bandwidth can be bought later. In two to three years, the costs will plummet such that the new requirements can be met with incremental or marginal costs.

It's wiser to buy bandwidth as you need it and not before (a small amount of incremental add-on will be applied, but it will be limited). The future will have

■ More choices

■ Increased providers

■ Greater availability

■ Lower costs

What should be done in the interim to satisfy the need for bandwidth? The answer is

■ Lease (dark) fiber instead of paying the cost of installing

■ Lease services from the ILEC or CLEC if sufficient bandwidth is available

■ Buy a wireless connection such as point-to-point microwave

In Table 2-1, a comparison of distances and frequencies is used for representative purposes. Many times this is the best-case scenario.

The myths run rampant with radio-based systems. Despite the rumors about the various risks and perils for the radio signal, microwave usually

Table 2-1

Comparison of Frequency Bands and Distances (Line of Sight)

Frequency Band	Distances (approximate)
2–6 GHz	30 miles
10–12 GHz	20 miles
18 GHz	7 miles
23 GHz	5 miles
28–30 GHz	1 to 2 miles

operates 99.99+ percent of the time. Microwave is normally impervious to the following:

- Snow
- Sleet
- Fog
- Birds
- Pollution
- Sandstorms
- Sunspot activity

The real risk is water fade (water absorption) and multipath fade across bodies of water. These can be accommodated for the most part in the design of the radio path.

A microwave link can transmit gigabytes of data without dropping a single bit (or packet when a data transmission uses packetized information). On copper wire, noise is always present. Thermal noise causes a continuous hum, white noise, and the like. A microwave path can be so clear that if no one is talking or sending data, the line is perfectly silent. This is difficult for the average layperson to understand.

Another perceived drawback (handling over a megabit per second) is the requirement of having line of sight between locations. Line-of-sight issues are rarely showstoppers. First, only a small part of the remote site needs to be visible. Even if the other site is not visible, solutions are available. Actually, this is quite common. Look for high points that can be used to get visibility between both sites. A passive or active repeater site can be implemented. Setting up a repeater is not difficult, particularly passive repeaters. The antennas are small and lightweight enough to be placed most anywhere: a water or radio tower, utility pole, other rooftops, and so on. An alternative is to bounce the signal off a physical obstacle (such as a mountain) and use obstacle gains to get the signal through. This is shown in Figure 2-9.

The 23-GHz frequency band, for example, is a very common frequency band for short-haul, private-user microwave systems. Most people confuse the band with the actual operating frequency. The 23-GHz band actually consists of 24 pairs of frequencies ranging between 21.200 and 23.600 GHz. The number can be doubled to 48 pairs of frequencies with minor antenna changes (changing the polarization from vertical to horizontal).

Figure 2-9
Obstacle gain uses a
bounced signal off a
natural obstacle such
as a mountain.

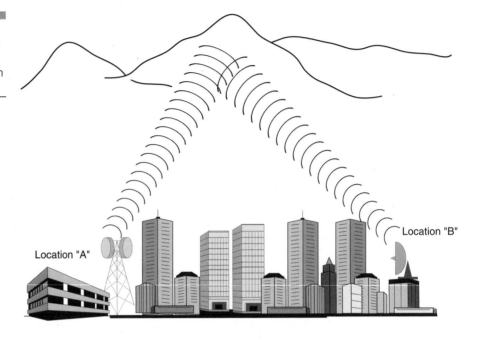

Location "A"

Location "B"

The radio signal is narrowly focused by the antennas at each end of the link, and transmit power is only about 60 milliwatts. These variables make it possible to use identical frequency pairs for two links originating from the same rooftop. By changing the polarization of the antennas and separating the signals, 10 degrees should provide the necessary separation and isolation.

Satellite Communications

Satellite communications have developed through the years. The first use of satellite technology was in the military for voice communications in the early 1960s. Despite the advancements in the technology, commercial providers are prohibited from constructing and launching satellites at will. Despite these limitations, commercial satellites provide valuable information on meteorology, agriculture, forestry, geology, environmental science, and other areas.

Countries use satellites in many adaptations. From military applications, agriculture, and oceanography, satellites have improved the overall performance of the individual country. Each of the applications listed, however, uses significant amounts of data to perform and monitor these functions.

Commercial Providers

NASA has been working with the U.S. commercial satellite industry to achieve seamless interoperability of satellite and terrestrial telecommunications networks. The high latency and noise characteristics of satellite links cause inefficiencies when using *Transmission Control Protocol* (TCP) in an environment for which it was not designed. These problems are not limited to airborne communications and may appear in other high latency and noisy networks. Considerations such as higher data rates and more users on data networks can affect future terrestrial networks similar to those posed by satellites today.

In the future, telecommunications networks with space stations, lunar stations, and planetary stations with their inherent latency, noise, and bandwidth characteristics need to be addressed. For now, the emphasis is on Earth-orbiting communications satellites.

How Do Satellites Work?

The basic elements of a satellite communications system are shown in Figure 2-10. The process begins at an earth station—an installation designed to transmit and receive signals from a satellite in orbit around the earth. Earth stations send information via high-powered, high-frequency (GHz range) signals to satellites, which receive and retransmit the signals back to the earth where they are received by other earth stations in the coverage area of the satellite. The area that receives a signal of useful strength from the satellite is known as the satellite's footprint. The transmission system from the earth station to the satellite is called the *uplink*, and the system from the satellite to the earth station is called the *downlink*.

Satellite Frequency Bands

The three most commonly used satellite frequency bands are the C-band, Ku-band, and Ka-band. C-band and Ku-band are the two most common frequency spectrums used by today's satellites. To help understand the

Figure 2-10
Satellite
communication
basics

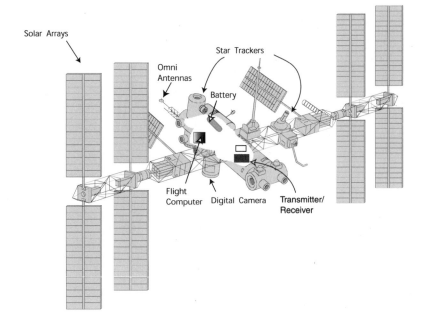

relationship between antenna diameter and transmission frequency, it is important to note that an inverse relationship exists between frequency and wavelength—when frequency increases, wavelength decreases. As wavelength increases, larger antennas are required to receive the signal.

C-band satellite transmissions occupy the 4- to 8-GHz frequency ranges. These relatively low frequencies translate to larger wavelengths than Ku-band or Ka-band. These larger wavelengths of the C-band mean that a larger satellite antenna is required to gather the minimum signal strength. The minimum size of an average C-band antenna is approximately 2 to 3 meters in diameter, as shown in Figure 2-11.

Ku-band satellite transmissions occupy the 11- to 17-GHz frequency ranges. These relatively high frequency transmissions correspond to shorter wavelengths, and therefore, a smaller antenna can be used to receive the minimum signal strength. Ku-band antennas can be as small as 18 inches in diameter. Figure 2-12 shows the Ku-band antenna of the Sony *Digital Satellite System* (DSS).

Ka-band satellite transmissions occupy the 20- to 30-GHz frequency range. These very high-frequency transmissions mean very small wavelengths and very small-diameter receiving antennas.

Figure 2-11
C-band satellite antenna

Figure 2-12
Ku-band satellite antenna

Geosynchronous Earth Orbit (GEO) Satellites Today, the over-whelming majority of satellites in orbit around the earth are positioned at a point 22,300 miles above the earth's equator in *geosynchronous earth orbit* (GEO). As shown in Figure 2-13, at a distance of 22,300 miles, a satellite can maintain an orbit with a period of rotation around the earth exactly equal to 24 hours. Because the satellites revolve at the same rotational speed of the earth, they appear stationary from the earth's surface. That's why most earth station antennas (satellite dishes) don't need to move once they have been properly aimed at a target satellite in the sky.

Mid-Earth Orbit (MEO) Satellites During the last few years, techno-logical innovations in space communications have given rise to new orbits and new system designs. *Mid-earth orbit* (MEO) satellite networks have

Figure 2-13
Geosynchronous
orbit

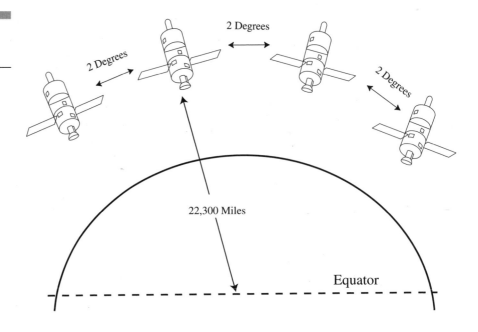

been proposed that will orbit at distances of about 8,000 miles. Signals transmitted from a MEO satellite travel a shorter distance, which translates to improved signal strength at the receiving end. This means that smaller, lighter-receiving terminals can be used. In addition, because the signal is traveling a shorter distance to and from the satellite, less transmission delay occurs. *Transmission delay* is the time it takes for a signal to travel up to a satellite and back down to a receiving station. For real-time communications, the shorter the transmission delays the better. For example, a GEO satellite requires .25 seconds for a round trip. A MEO satellite requires less than .1 seconds to complete the job. MEO operate in the 2-GHz and above frequency range.

Low-Earth Orbit (LEO) Satellites   Proposed LEO satellites are divided into three categories: little LEOs, big LEOs, and Mega-LEOs. LEO will orbit at a distance of only 500 to 1,000 miles above the earth. This relatively short distance reduces transmission delay to only .05 seconds and further reduces the need for sensitive and bulky receiving equipment. Little LEOs will operate in the 800 MHz (.8 GHz) range, big LEOs will operate in the 2-GHz or above range, and Mega-LEOs will operate in the 20- to 30-GHz range. The higher frequencies associated with Mega-LEOs translate into more information-carrying capacity and the capability of real-time, low-delay video transmission. Microsoft Corporation and McCaw Cellular (now known as AT&T Wireless Services) have partnered to deploy 288 satellites to form Teledesic, a proposed Mega-LEO satellite network. Speeds are expected at 64 Mbps downlink and 2 Mbps uplink.

Orbital Slots

With more than 200 satellites in geosynchronous orbit, how do we keep them from running into each other or from attempting to use the same location in space? To tackle that problem, international regulatory bodies like the *International Telecommunication Union* (ITU) and national government organizations like the FCC designate the locations on the geosynchronous orbit where communications satellites can be located. These locations are specified in degrees of longitude and are known as orbital slots. In response to the huge demand for orbital slots, the FCC and ITU have progressively reduced the required spacing down to only 2 degrees for C-band and Ku-band satellites.

Communications

The communications subsystem of a satellite is essential to the function of all satellites. Many different components are used in space borne communications subsystems including

- Special antennas
- Receivers
- Transmitters

All components must be highly reliable and low weight. Most satellites are fitted with beacons or transponders, which help with easy ground tracking. Global receive and transmit horns receive and transmit signals over wide areas on the earth. Many satellites and ground stations have radio dishes that transmit and receive signals to communicate. The curved dishes reflect outgoing signals from the central horn and reflect incoming signals. A satellite's capability to receive signals is also necessary in order to trigger the return of data or to correct a malfunction if possible.

The technological and regulatory hurdles to create true high-speed satellite networks are quickly becoming past tense. Low- and mid-bandwidth systems such as Motorola's Iridium and Hughes' DirecPC can handle some of the needs immediately. These systems are nothing compared to the promise of 2 Mbps, 20 Mbps, and even 155 Mbps streaming down from the sky. All that is necessary is a small antenna, a black box, and a service provider. This will approximate the way we buy service from an *Internet Service Provider* (ISP) today.

Is it ready for prime time yet? Not yet! Iridium's universal telephone didn't kill the cellular telephone. Therefore, broadband satellite systems will not kill terrestrial lines. Broadband satellite creators agree that broadband satellite systems will complement terrestrial networks. These satellites will provide high-speed service where terrestrial infrastructure does not exist. However, high-speed, low-cost landlines are here to stay.

Is there an application for the high-speed satellite networks? What makes them different from each other? Each of the main systems is very different. Some of the most visible ones may prove the most difficult to implement. Some of the most staid-looking systems may beat every other system to the punch.

Satellite communications is nothing new. For years, you could hook up a *very small aperture terminal* (VSAT) system and buy time on a satellite. The VSAT can deliver up to 24 Mbps in a point-to-multipoint link (a multicast) and up to 1.5 Mbps in a point-to-point link. These systems are fine for predictable and quantifiable data transmission, but if you need to conduct business on the fly, they can be a problem. New techniques are required to handle the on-demand usage. Primary among them are more tightly focused beams and digital signal technology, which together can increase frequency reuse (and thereby increase bandwidth) and reduce dish size from meters to centimeters. Accordingly, you also need a large unused chunk of the electromagnetic spectrum.

In 1993, NASA launched its *Advanced Communication Technology Satellite* (ACTS). ACTS is an all-digital, Ka-band (20–30 GHz), spot-beam, GEO satellite system capable of delivering hundreds of megabits per second of bandwidth. NASA showed that such a system could work. The FCC has since granted orbital locations and Ka-band licenses to 13 companies including

- EchoStar
- Hughes
- Loral
- Motorola
- Ka-Star
- NetSat 28

- PanAmSat
- Teledesic

They all aim to bring bandwidth up to 155 Mbps to our home and office. What remains to be seen is how many will actually launch and how many will survive.

What will we do with this speed and capacity? Anything we want! Whatever landlines can do, the satellite systems of the new millennium will also do. Table 2-2 lists the applications covered by broadband satellite communications. These are representative applications; others will be developed.

Most of the market that needs data services seems to be well served by landlines. So, why is the emphasis on the use of airborne technology? An obvious market is in places that have underdeveloped communications infrastructures. In some countries, stringing copper or fiber is out of the question because of the initial cost and the terrain where it is needed. Still, a wireless telephone has some merit.

So, who *does* need this new class of broadband satellite communications? The first answer is the same: multinational corporations. The main problem satellite systems solve is getting high-bandwidth access to places without a high-bandwidth infrastructure. It's unlikely that a satellite system could compete with *Digital Subscriber Line* (xDSL) to the home or fiber to the office.

Table 2-2

Applications for High-Speed Satellite Communications

Applications
Desk-to-desk communications
Videoconferencing
High-speed Internet access
E-mail
Digital and color fax
Telemedicine
Direct TV and video
Transaction processing
Interactive distance learning
News and financial information
Teleradiology

LEO versus GEO

However, bandwidth is only half the story. The other half is latency. It's easy to talk about high-bandwidth satellite systems, but that technology has existed in VSATs for years. GEO satellites located at 22,300 miles above the equator induce 250 milliseconds of round-trip delay. With that kind of latency built into the system, (not counting latency added by the various gateways and other translations), a telephone conversation is annoying. Any interactive data-intense application has to be nonlatency-sensitive. Online transaction processing will have a problem using a GEO satellite system.

Moving the satellites closer to the earth will help significantly. That's just what systems such as Teledesic[1], Skybridge[2], and Celestri[3] will do. With LEOs under 1,000 miles, these systems reduce latency to .1 second. Whereas GEOs are a well-known technology, LEOs are new. The biggest problem is that you need many of them to have global coverage. At one point, Teledesic planned a constellation of more than 842 satellites.[4] Until recently, the concept of launching dozens or hundreds of multimillion-dollar satellites was a pipe dream. Estimates for each of Teledesic's 288 satellites are $20 million apiece.

Price is only one issue. Finding a company to launch all these satellites poses another obstacle. Teledesic set an 18-month to 2-year launch window to get its 288 satellites airborne. LEO system planners are talking about putting more satellites into orbit in the next five years than the world has put into orbit over the past 40 years. Once the LEO satellites are in orbit, an entirely new set of problems arises; the matter of space junk becomes a problem.

Niches in the GEO Sphere

LEOs will be good for high-speed networking, teleconferencing, telemedicine, and interactive applications. GEOs will be better for information downloading and video distribution such as broadcasting and multicasting.

---

[1] Teledesic is a joint venture between Microsoft and McCaw Cellular.

[2] Skybridge is a venture of Alcatel of Belgium.

[3] Celestri is a product and venture of Motorola.

[4] Teledesic dropped the number of required LEO satellites to 288 since the first announcement.

We're able to use GEO satellites to transport at least 24 Mbps of broadcast IP data and over 2 Mbps of point-to-point TCP/IP data. The latter uses technologies such as TCP spoofing. Several vendors have been using this technique for years to deliver Internet and Intranet content at high speed. Ground terminals can use similar TCP spoofing technologies. Nevertheless, the 250-millisecond delay that you just can't get around is still present. Any lossless protocol is going to have problems with this latency. Even if TCP spoofing works given TCP's 64-KB buffer make this somewhat suspect, other protocols such as IBM's SNA and other real-time protocols are designed around landline performance.

LEO Meets GEO

Motorola's Celestri plans an initial LEO constellation of 63 satellites coupled with one GEO satellite over the U.S. The LEO constellation and the GEO satellites will be able to communicate directly through a satellite-to-satellite network.

The hybrid configuration will enable Celestri to take advantage of LEO's shorter delays for interactive uses and GEO's power in the broadcast arena.

Space Security Unit

Once you get beyond the latency and bandwidth issues, another challenge arises: security. If data is being packaged up and broadcast into space, anybody with a scanner can just tune in. The air interface technologies that these systems use will make it more difficult for anyone to eavesdrop. Combined systems will use *code division multiple access* (CDMA), *time division multiple access* (TDMA), *frequency division multiple access* (FDMA), and several other xDMA protocols. On top of that, many of the networks will offer some internal security encryption system.

The Market for the Network

The total global telecommunications market is about $750 billion, and this will double in 10 years, chiefly due to expanded use of data communications. We'll use whatever service we have available to meet that demand: fiber, ATM, *Synchronous Optical Network* (SONET), xDSL,

Gigabit Ethernet, cable modems, satellites, and probably a few that haven't even been thought of yet.

All telecommunications systems will compete on availability, price, and speed. That means there will be two big winners: whoever gets its broadband service to consumers first and whoever can offer the most bandwidth with the least reasonable latency. Table 2-3 shows a sample of the techniques and the services being offered in the 2000 timeframe.

The bands are continually referred to in the discussion of the various bands that these services will operate in. As a means of showing the bands and the frequency they fall into, Table 2-4 provides a summary of some of the bands used in the satellite arena.

Table 2-5 shows a sample of the satellite technologies that use the various bands and the size of the terminal and dish in satellite technologies.

Table 2-3

Sampling of
Services Planned

Service Offering	Astrolink	Teledesic	Celestri
Companies (parties) involved	Lockheed	Bill Gates, Craig McCaw, Boeing	Motorola
Service offerings	Data, video, rural telephony	Voice, data, video-conferencing	Voice, data, video-conferencing
Satellite orbit	22,300	435	875 and 22,300
Satellite band	Ka	Ka	Ka and also 40–50 GHz
Dish size	33–47 inches	10 inches	24 inches
Bandwidth	Up to 9.6 Mbps	16 Kbps–64 Mbps or up to 2.048 Mbps	Up to 155 Mbps transmit and receive
Available (approx.)	Late 2000	2002	2002
Number satellites	9 GEOs	288 LEOs	63 LEOs, up to 9 GEOs
Access technologies	FDMA, TDMA	MF-TDMA, ATDMA	FDMA, TDMA
Inter-satellite Communications	Yes	Yes	Yes

Table 2-4

Sample of the
Frequency Bands

Band	Frequency Ranges Used
HF-band	1.8–430 MHz
VHF-band	50–146 MHz
P-band	0.230–1.000 GHz
UHF-band	0.430–1.300 GHz
L-band	1.530–2.700 GHz
FCC Digital Radio and PCS	2.310–2. 360 GHz
S-band	2.700–3.500 GHz
C-band	Downlink: 3.700–4.200 GHz Uplink: 5.925–6.425 GHz
X-band	Downlink: 7.250–7.745 GHz Uplink: 7.900–8.395 GHz
Ku-band (Europe)	Downlink: FSS: 10.700–11.700 GHz DBS: 11.700–12.500 GHz Telecom: 12.500–12.750 GHz Uplink: FSS and Telecom: 14.000–14.800 GHz; DBS: 17.300–18.100 GHz
Ku-band (America)	Downlink: FSS: 11.700–12.200 GHz DBS: 12.200–12.700 GHz Uplink: FSS: 14.000–14.500 GHz DBS: 17.300–17.800 GHz
Ka-band	Roughly 18–31 GHz

Satellite Characteristics

As already mentioned, satellite communications have three general characteristics that lead to interoperability problems with systems that have not been designed to accommodate them:

- Delay (or latency)
- Noise
- Limited bandwidth

Table 2-5

Various Choices of Satellite Technologies Used

System Type	Frequency Bands	Applications	Terminal Type/Size
Fixed satellite service	C and Ku	Video delivery, VSAT, news gathering, telephony	1-meter and larger fixed earth station
Direct broadcast	Ku	Direct-to-home video/audio satellite	0.3–0.6-meter fixed earth station
Mobile satellite (GEO)	L and S	Voice and low-speed data to mobile terminals	Laptop computer/ antenna-mounted but mobile
Big LEO	L and S	Cellular telephony, data, paging	Cellular phone and pagers; fixed phone booth
Little LEO	P and below	Position location, tracking, messaging	6 inches; omnidirectional
Broadband GEO	Ka and Ku	Internet access, voice, video, data	20-cm, fixed
Broadband LEO	Ka and Ku	Internet access, voice, video, data, videoconferencing	Dual 20-cm tracking antennas, fixed

It is important to understand each of these in general terms and in more detail. Since the invention and installation of fiber optics, communications systems have been treated as ideal with very low latency, no noise, and nearly infinite bandwidth (the capacities are rapidly moving to terabit bandwidth from current fiber systems). These characteristics still have room for doubt and development, but for satellite communications, the characteristics of fiber can make it very difficult to provide cost-effective interoperability with land-based systems. The latency problem is foremost among the three. Noise can be handled by the application of error control coding. Bandwidth efficiency is an important goal for satellite systems today, but will become increasingly important as the number of users as well as the data requirements increase.

Latency

We have seen that an inherent delay occurs in the delivery of a message over a satellite link due to the finite speed of light and the altitude of communications satellites. As we stated, an approximately 250-millisecond propagation delay occurs in a GEO. These delays are for one ground station-to-satellite-to-ground station route (or hop). The round-trip propagation delay for a message and its reply would be a maximum of 500 milliseconds. The delay will be proportionately longer if the link includes multiple hops or if intersatellite links are used. As satellites become more complex and include on-board processing of signals, additional delay may be added.

Other orbits are possible including low-earth orbit (LEO) and mid-earth orbit (MEO). The advantage of GEO is that the satellite remains stationary over one point of the earth allowing the simple pointing of large antennas. The lower orbits require the use of constellations of satellites for constant coverage and are more likely to use intersatellite links, which result in variable path delay depending on routing through the network.

Noise

The strength of a radio signal falls in proportion to the square of the distance traveled. For a satellite link, this distance is very large, so the signal becomes very weak. This results in a low signal-to-noise ratio. Typical bit error rates for a satellite link might be on the order of 10^{-7}. Noise becomes less of a problem as error control coding is used. Error-performance equal to fiber is possible with proper error control coding.

Bandwidth

The radio spectrum is a finite resource and only so much bandwidth is available. Typical carrier frequencies for current point-to-point (fixed) satellite services 6/4 GHz (C-band); 14/12 GHz (Ku-band); and 30/20 GHz (Ka-band). Traditional C- and Ku-band transponder bandwidth is typically 36 MHz to accommodate one color television channel (or 1,200 voice channels). One satellite may carry two-dozen or more transponders.

New applications, such as personal communications services, may change this picture. Bandwidth is limited by nature, but the allocations for commercial communications are limited by international agreements so that the scarce resource can be used fairly. Applications that are not bandwidth efficient waste a valuable resource, especially in the case of satellites. Scaleable technologies will be more important in the future.

Advantages

Although satellites have certain disadvantages, they also have certain advantages over terrestrial communications systems. Satellites have a natural broadcast capability. Satellites are wireless, so they can reach geographically remote areas. Major changes are occurring in the design of communications satellite systems. New bandwidth-on-demand services are being proposed for GEO. The Federal Communications Commission received 17 filings for new Ka-band satellite systems last year.

Low-Earth Orbit Satellites (LEOs)

Quite a bit of discussion regarding wireless communications has already been covered. However, a combination of two separate services and technologies are merging as new services for broadband communications. These two services include long-haul communications and the use of personal communications services. The technologies include the use of satellite and the cellular concepts combined. Worldwide communications services can be achieved by these two combined services, therefore some diligent effort should be used to understand just what is happening in this arena. This truly brings home the concept of communications from anywhere to anywhere. The thought of being out in the middle of a lake and receiving a call or rafting down a river and making a call boggles the mind. This is especially true when thinking about some of the most rural areas in the world where no telephone service infrastructure exists today. Yet, in a matter of a few years these remote locations, on mountaintops, in forests, in valleys, and on sea will all be reachable within a moment's notice. The infrastructure of a wired world will not easily lend itself to this need due to timing and cost issues. Therefore, the use of a wireless transmission system is the obvious answer. However, the use of cellular and personal communications devices still leaves much to be desired. First, the deployment of these services is always going to be in the major metropolitan areas, where the use and financial payback will be achieved. Thus, in the remote areas, it will be decades if not longer, until the deployment ever works its way close to the remote areas. Enter the ability to see the world from above the skyline! The industry decided to attempt servicing remote areas from a satellite capacity. This is not a new concept; the use of satellite transmission systems has been around for over thirty years. However, the application for an on-demand, dial-up satellite service is new. This would have to be a lucrative business venture because the costs are still quite high. Look at the Iridium project that took over five years to build and launch, only to meet with lax reception from the marketplace. This drove Iridium Inc. into bankruptcy in their first year of operation. Approximately one dozen suppliers are still competing for space segment and frequency allocation to offer voice, broadband data, paging, and determination services. In each case, the organizations selected various approaches on how to launch their service offerings and the use of orbital capabilities to provide the service. In general, this discussion will cover the most widely discussed concept today called low-earth orbit (LEO). However, some added discussion on the use of mid-earth orbit (MEO) and geosynchronous orbit (GEO) satellite-based communications offerings and applications must be addressed. As mentioned earlier, a dozen organizations are applying for the rights to use various orbital slots and frequencies to provide service in the various orbits, as shown in Table 2-6. The licenses requested are being discussed at great length around the world as the future interoperable service.

Low-Earth Orbit

In December 1990, Motorola filed an application with the FCC for the purposes of constructing, launching, and operating a LEO global mobile satellite system known as Iridium. This hot button sparked the world into frenzy. Iridium was a concept of launching a series of 66 satellites[5] around the world to provide global coverage for a mobile communications service

Table 2-6

A Summary of the Number of Competitors and the Various Orbits Being Sought

Orbit	Number of Competitors	Status
Low-Earth Orbit (LEO)	8	Pending licenses granted based on very specific areas of coverage.
Mid-Earth Orbit (MEO)	4	Experimental licenses granted for specific areas of coverage.
Geosynchronous Orbit (GEO)	4	Licenses have already been issued for some, some are experimental with others.

[5] Originally, the Iridium proposal was for 77 satellites, but Motorola amended this number after the World Administrative Radio Council meeting in the spring of 1992.

operating in the 1.610- to 1.6265-GHz frequency bands. The concept is to use a portable or mobile transceiver with low-profile antennas to reach a constellation of 66 satellites. Each of the satellites is interconnected to one another through a radio communications system as they traverse the globe at 413 nautical miles above the earth in multiple polar orbits.[6] This provides a continuous line-of-sight coverage area from any point on the globe to virtually any other point on the globe, using a spot beam from the radio communications services on board each of the satellites. The use of this spot beam concept, which has been discussed for years in the satellite industry, enables high frequency reuse capacities that had not been achieved before. Iridium wanted to provide the services outlined in Table 2-7. The concept is sound and the approach provides the coverage that was lacking in the past to remote areas. In the table, the two columns are used as exclusive of each other. The services can be provided in any of the coverage areas regardless of which service is selected.

Each of the satellites is relatively small in terms of others that have been used. The electronics inside each is very sophisticated and the use of gateway controllers on Earth provides the command and control service for the administration of the overall network. With the Iridium network, cellular communications is expanded providing coverage in areas where cell sites would not be practical. This is summarized in Table 2-8, a feature comparison of the cellular concept versus the Iridium concept.

Figure 2-14 shows the concept of the LEO arrangement. In this particular case, the satellites are traversing the earth's surface at a height of 400+ nautical miles above the earth in a polar orbit. In the polar orbit, the satellite moves around the earth's poles and passes over any specific point along

Table 2-7

Services and Coverage Through the Iridium Network

Type of Service	Coverage Areas
Voice communications	Air
Data communications	Land
Paging	Water
Radio determination services	

---

[6] The original concept was to use 7 polar orbits with 11 satellites in each. This would provide worldwide coverage, much similar to an orange slice concept.

Table 2-8

Summary of Cell Sites Versus Iridium Cells

Cellular Networks	Iridium Network
Sites are fixed.	Sites are the moving targets.
Users move from site to site.	User stays put, sites move the user from satellite to satellite.
Areas of coverage are 3 to 5 miles across.	Areas of coverage are 185–1,100± miles across.
Coverage sporadic, not totally ubiquitous.	Worldwide coverage.

Figure 2-14
The LEO concept

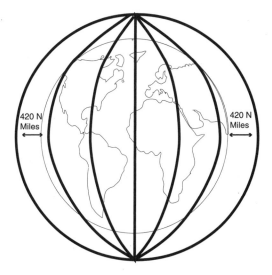

its path very quickly. The satellites move at approximately 7,400 meters ± per second in different orbits. Therefore, as one target site moves out of view, a new one comes into view at approximately the same time. A handoff will take place between the individual satellites (using the Ka-band).

Figure 2-15 represents the ground telemetry and control services, called gateway feeder links. These also use spectrum in the Ka-band. Iridium used approximately 16.5 MHz of bandwidth in the L-band. The L-band is also used from the handset to the satellite, whereas the Ka-band is used from satellite-to-satellite communications, as shown in Figure 2-16. The use of this L-band enables low-powered handsets to communicate within the 413 nautical mile distances with the satellites.

Figure 2-15
Ground station
telemetry and
control

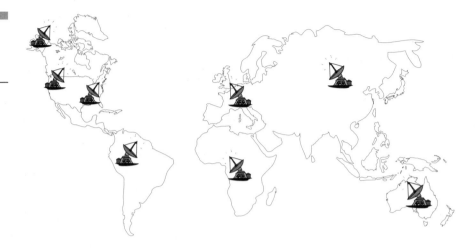

Figure 2-16
Intercommunications
between satellites
with Ka-band

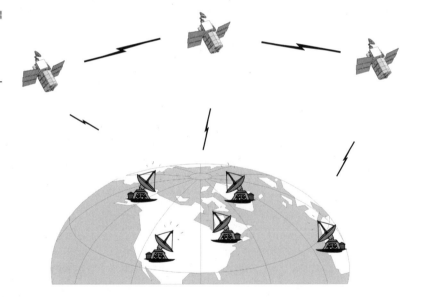

The Benefits of These Service Offerings

Generically, these benefits are addressed and kept in the context of any LEO network. The benefits lean toward the end user, as shown in the following list:

- **Ubiquitous services** With continuous and global coverage, any to all connections can occur. As users travel either domestically or abroad, the service travels with them. It will eliminate the need for special access arrangements and special numbers that must be dialed. The user should never be out of range from his or her network. Remote areas with limited demand and finances now have the capability to connect anywhere in the world.

- **Spectral efficiency** As already mentioned, the frequency reuse patterns for the bandwidth allocation will be significant. No other satellite system has achieved these reuse ratios. Iridium was first to claim this capability of efficiency. The RDSS portion of the Iridium network is contained in the same spectral arrangement, freeing up 16.5 MHz of spectrum. This is a quantum leap in the efficient use of the spectrum.

- **Public benefits due to flexible design** The digital technology deployed enables the total connection for all voice and data services on a 7-day by 24-hour basis. This allows the flexibility of service provisioning. The low-earth orbit overcomes some of the limitations of the higher transport systems, such as the delay in the round-trip transmission. Because the satellites are low, the user set needs a lower power output device. This orbit has been selected to be the most flexible.

- **The potential to save lives** How often has the news media published stories of people stranded in remote areas with no life support systems and therefore died? The press today is filled with stories of cellular and PCS users notifying authorities of casualties. If only people in remote areas had a means of notifying authorities and/rescue parties, their lives may have been spared.

- **Capabilities of the vendor** Motorola states that they are uniquely qualified to provide this form of services due to their background in the development and sales of other ancillary equipment that works in the wireless world. Specifically, they have been one of the major developers in the production, research, and development of private mobile-radio services.

- **LEO deployment promotes international communications** The LEO networks deliver modern digital-transmission services to remote areas of the world. The FCC and the U.S. government are attempting to use telecommunications as a strategic and economic tool to foster development in these areas. Their goals are to

  - Promote the free flow of information worldwide

- Promote the development of innovative, efficient, and cost-effective international communications services that meet the needs of users in support of commerce and trade development

- Continuous development and evolution of a communications service and network that can meet the needs of all nations, and specifically those of developing nations

These goals can be met with a mobile communications network such as the one proposed in the LEO networks. An alternative is the *global services mobile* (GSM) standard that is emerging throughout the international arena.

The LEO satellites are spaced at 32.7 degrees apart, travel in the same basic direction moving at approximately 16,700 miles per hour from north to south and 900 miles per hour westward over the equator. Given this path, each satellite is designed to circle the earth at approximately every 100 minutes. At the equator, a single device will provide coverage. However, as the craft moves toward the poles, overlap will occur increasing levels of coverage above and below the equator. The expected life cycle of the satellites is around five years.

The basic building blocks of the original Iridium system provided the model for other networks such as Globalstar systems. In each case, the systems use proven technology for radio transmission in well-established frequency bands. The basic system is composed of the following:

- The space segment comprised of the constantly moving constellation of satellites in a low-earth orbit

- A gateway segment comprised of earth station facilities around the world

- A centralized system control facility

- The launch segment to place the craft in the appropriate orbit

- A subscriber unit to provide the services to the end user

In 2000, Iridium, Inc. filed bankruptcy because prices for the sets and the service were too high. Motorola looked for a suitor and in 2001 Iridium, LLC, a new company emerged. This preserved the system and prevented the de-construction of the network. Globalstar experienced similar financial problems.

Global Positioning System

The *Global Positioning System* (GPS) was developed for the U.S. military, but used to provide positional information for commercial and even leisure applications. The GPS system consists of a bracelet of satellites transmitting information about their position relative to the earth and very accurate timing information. A small receiver in a vehicle can determine its position on the surface of the earth by receiving signals from at least three satellites. With three satellites, the position can be determined in two dimensions, but with four or more signals received, the altitude can be measured as well. In open country, a receiver can normally receive information from five satellites. The positional information can be calculated by the receiver knowing how long the radio signal takes to reach it from each satellite (and thus its distance from it) and the position of each satellite in space. The U.S. military has built in a random error mechanism so that other users cannot achieve the same accuracy as official users, who access the GPS information on a separate encrypted radio channel.

GPS equipment has now been developed that is highly miniaturized and ruggedized and may be carried by people who are walking in remote areas or by small boat sailors. The cost of these devices has reduced to the point where they are no more expensive than a good quality VCR.

Direct Broadcast Satellites

The satellite systems that are most familiar to us are the *Direct Broadcast Satellite* (DBS) systems. These can carry hundreds of TV channels in addition to other services. DBS satellites are normally in geostationary orbit, which means they appear to be in one position in the sky at all times. This is achieved by placing them in an orbit whose period is identical to that of the earth's rotation. The altitude required for a geostationary circular orbit is 35,803 km.

Communication Satellites

Communication satellite systems are used to access remote parts of the world, as well as providing intercontinental telephone, TV, and data links. The intercontinental services are provided by fixed ground stations in each continent and form part of the international telecommunications network. Private satellite service operators such as Inmarsat and Eutelsat provide communication services to mobile ground stations. The mobile ground stations tend to be bulky and have to have their antennae accurately positioned to establish communications. These services include TV and voice reporting from remote locations as well as data services.

Companies such as IBM can offer communication satellite uplink and downlink services, providing communication satellite access without the need for the customer to invest in his or her own ground stations.

Other Commercial Applications

The ability to access strategically important and mission-critical applications is vital to many companies. There is a need to extend communications beyond landlines. Time is a critical component. For example, a parcel delivery service must be able to redirect delivery vans at a moment's notice. Many wireless applications are already in operation:

- Radio and TV stations
- Telecommunications links
- Some utility companies can now remotely read gas or electricity meters from a vehicle parked in the street outside.
- Satellite communications
  - Voice links
  - Data links
  - Video broadcasting
  - Taxi communication
- Military vehicle communications
- Surveillance equipment
- Wireless mouse/keyboard connections for the PC
- Wireless local area networks (LANs)

Wireless data communication has been around for a very long time. When we talk about "wireless" in this book, we mean untethered, and when we say data, we mean any communication method other than speech. "Radio" means radio frequency (RF), whereas wireless includes RF, infrared, ultrasonic, and other technologies. Wireless communications will make a significant impact to the information technology industry in the next few years and many new applications will evolve as a result.

Some of the main applications of wireless technology are

- Early voice communication systems
- Analog cellular telephones
- Digital cellular telephones
- Cordless phones
- Paging services
- *Private Mobile Radio* (PMR and PAMR)
- Mobile data services
- Satellite communications
- Other domestic and commercial applications

Access
Techniques for
Radio-Based
Systems

Some people say that getting there is most of the work. The same holds true with the means of getting either voice or data ready for transmission. Several techniques are available to prepare the information, including the modulation and the access methods. Moreover, the use of simplex or duplex channel capacities plays into the overall structure of the wireless networks. Whereas a voice channel is automatically set up in the telephony world to be full duplex (even if we only use it for a half-duplex operation), the radio world uses the channel differently. We shall analyze these differences in this chapter, not from a selection of better or worse, but from a perspective of the choices available.

Frequency Division Multiple Access (FDMA)

Frequency Division Multiple Access (FDMA) is used for standard analog cellular radio systems. Each user is assigned a discreet slice of the RF spectrum. Therefore, only one user per channel is permitted because it enables the user to occupy the channel 100 percent of the time. As shown in Figure 3-1, the slices vary depending on the standard being used.

Figure 3-1
FDMA channels permit one user per narrowband channel.

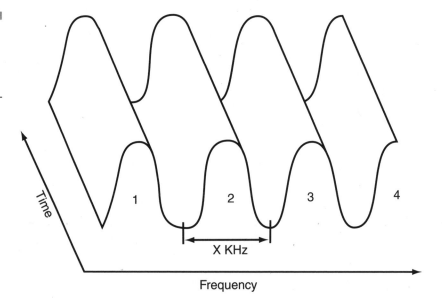

Because FDMA uses the full 30-kHz channels for one telephone call at a time, it is obviously wasteful. FDMA, an analog technique, can be improved a little by using the same frequency in a *Time Division Duplex* (TDD) mechanism. In this mechanism, one channel is used and time slots are created. The conversation flows from A to B and then from B to A. The use of this channel is slightly more efficient. However, when nothing is being sent, the channel remains idle. Because digital transmission introduces better multiplexing schemes, the carriers want to get more users on an already strained radio frequency spectrum. See Figure 3-2 for a layout of the FDMA architecture.

Additional possibilities for enhancing security and reducing fraud can be addressed with digital cellular and *personal communications services* (PCS). Again, this appears to be a win-win for the carrier:

- Less costs

- More users

- Better security

- Less fraud

Obviously, the carriers welcomed some of these discussions.

Once the decision was made to consider digital transmission, the major problem was how, what flavor to use, and how to seamlessly migrate the existing customer base to digital.

Figure 3-2
FDMA slotting

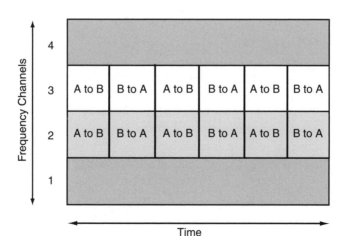

The digital techniques available to the carriers are

- *Time Division Multiple Access* (TDMA)
- *Extended Time Division Multiple Access* (ETDMA)
- *Code Division Multiple Access* (CDMA)
- *Global Services Mobile* (GSM) (using a slightly different form of TDMA)
- *Narrowband Advanced Mobile Phone Service* (N-AMPS)

Carriers and manufacturers are using each of these systems. The various means of implementing these systems have brought about several discussions regarding the benefits and losses of each choice.

Digital Cellular Evolution

As the radio spectrum for cellular and PCS continues to become more congested, the two primary approaches in North America use derivatives of

- TDMA
- CDMA

Each move to digital requires newer equipment, which means capital investments for the PCS carriers. Moreover, as these carriers compete for radio frequency spectrum, they have to make significant investments during an auction from the FCC. This places an immense financial burden on these carriers before they even begin the construction of their networks.

Time Division Multiple Access (TDMA)

TDMA uses a time division-multiplexing scheme where time slices are allocated to multiple conversations. Multiple users share a single 30-kHz radio channel without interfering with each other because they are kept separate by using fixed time slots, as shown in Figure 3-3.

The evolution of TDMA has been fairly quick to get to a higher reliability and denser occupancy of the spectrum. Figure 3-4 is a time line of the evolution of the various standards for TDMA.

The current standard for TDMA divides a single channel into six time slots and then three different conversations use the time slots by allocating

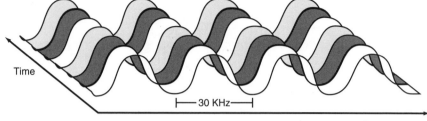

Figure 3-3
IS-54 standards for TDMA share the channel among three users.

Time

├── 30 KHz ──┤

Frequency

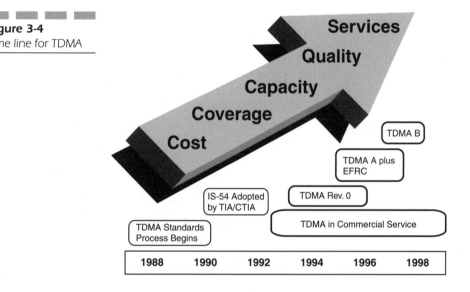

Figure 3-4
Time line for TDMA

Services
Quality
Capacity
Coverage
Cost

TDMA B

TDMA A plus EFRC

IS-54 Adopted by TIA/CTIA

TDMA Rev. 0

TDMA in Commercial Service

TDMA Standards Process Begins

| 1988 | 1990 | 1992 | 1994 | 1996 | 1998 |

two slots per conversation. This provides a threefold increase in the number of users on the same radio frequency spectrum. The industry was looking for a ten-fold increase in capacity, but TDMA in the current state produced only a threefold increase. Although TDMA deals typically with an analog-to-digital conversion using a typical pulse code modulation technique, it performs differently in a radio transmission. *Pulse code modulation* (PCM) is translated into a quadrature phase-shift keying technique thereby producing a four-phased shift, doubling the data rate for data transmission. The typical time slotting mechanism for TDMA is shown in Figure 3-5.

The wireless industry began to deploy the use of TDMA in the early 1990s when the scarce radio frequency spectrum problem became most noticeable. The intent was to improve the quality of radio transmission as well as the efficiency usage of the limited spectrum available. TDMA can

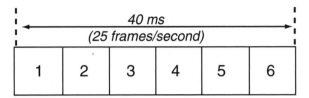

correctly increase the number of users on the RF by three times. The three-fold increase was enough to get the carriers up and running quickly and held some of the costs to a reasonable level. However, as more enhanced PCS techniques such as micro- and picocells are used, the number can grow to as much as a 40-fold increase in the number of users on the same RF spectrum. One can see why it is so popular.

TDMA has another advantage over the older analog (FDMA) techniques. Where the analog transmission across the 800-MHz frequency band (in North America it is 800 MHz, in much of the world it is 900 MHz) supports one primary service, that being voice, the TDMA architecture uses a PCM input to the RF spectrum. Therefore, TDMA can also support digital services for data in increments of 64 Kbps. The data rate can support from 64 Kbps to hundreds of megabits/second (120 Mbps).

Carriers like TDMA because of its capability to add data and voice across the RF spectrum and the cost associated with the migration from analog to digital. This is an attractive opportunity for the carriers who are developing PCS and digital cellular using the industry standards. The two standards in use today include IS-54, which is the first evolution to TDMA from FDMA and IS-136, the latest and greatest technology for 800-900 and 1,900 MHz TDMA service. When the IS-136 service was introduced, the addition of data and other services were also introduced. These include the *short message service* (SMS), caller-ID display, data transmission, and other service levels. Using a TDMA approach, carriers feel that they can meet the needs for voice, data, and video integration for the future. TDMA is the basis of the architecture for the GSM standard in Europe and Japan. Although TDMA, GSM uses a different framing format, which is shown in Figure 3-6.

IS-136 TDMA

The motivation for spectrum owners to make plans to use up banded second-generation cellular technologies for use in the PCS bands in the

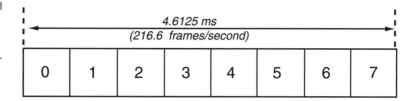

Figure 3-6
GSM/TDMA time
slotting

U.S. takes on many considerations. Second-generation cellular, however, will need some significant improvements. The evolution to IS-136 TDMA will be the enabling technology to accentuate and deliver the features and data communications capabilities that users await. Much higher voice communications and in-building modes of operation become the norm with the use of newer digital cellular transmission. High-speed packet data transmissions are also on the horizon for TDMA with IS-136 standards implementations.

The IS-136 TDMA standard creates the opportunity for integrating macrocellular and indoor wireless access in a seamless fashion while using a low-cost handset that is based entirely on 30-kHz frequency division duplex channels. The IS-136 standard includes provisions for private system IDs and for the integration of private and public wireless access. Because IS-136 is based on narrowband 30-kHz channels with hundreds of channels in a commercial system, schemes that borrow channels from a public system for the operation of a low-power private indoor system can work well. Private systems can borrow spectrum in small pieces and they generally can find quiet spectrum while operating within a macrocellular system. Studies of *personal base stations* (PBS) operating within a macrocellular system using shared spectrum have shown that user densities can be supported that are as high as 300 to 1,250 PBS per square mile.

The framing of the TDMA IS-136 format is shown in Figure 3-7 whereby a 40-ms frame will be used to create the data channel and possibly give us an across-the-air interface at 48.6 Kbps for now.

Capabilities Provided Through TDMA

Seamless system transition is available with TDMA between private and public systems where a need exists to interwork between the various operators' systems and the independent (private) system in a building. Differentiated charging is also a possibility based on the public and private operation that is being used by the subscriber.

Figure 3-7
The frame for TDMA
IS-136

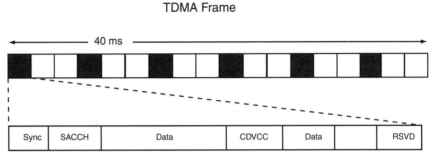

Alphanumeric display of a serving system can be provided at the user's handset so that the user can differentiate how and where to use the system. TDMA becomes an underlay technology and in-building support capability for both private or semi-private systems operations. It is also possible to define multiple private systems off a single *digital control channel* (DCCH).

The digital control channel can support both forward and reverse control channel services, as shown in Figure 3-8. The necessary channel control functions are shown in this figure.

Looking even further into the framing and the control channels, Figure 3-9 provides a sample of the forward and reverse digital control chan-

Figure 3-8
The digital control
channel functions

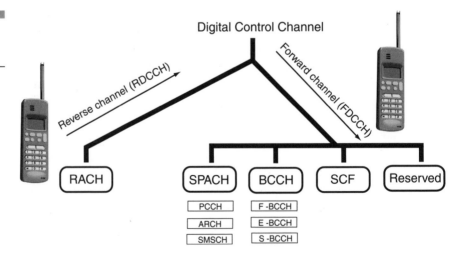

Figure 3-9
Control channels in
the TDMA framed
format

Forward DCCH

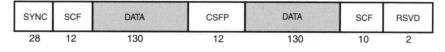

SYNC	SCF	DATA	CSFP	DATA	SCF	RSVD
28	12	130	12	130	10	2

Forward DTCH

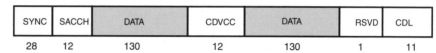

SYNC	SACCH	DATA	CDVCC	DATA	RSVD	CDL
28	12	130	12	130	1	11

Subfields

nels that use the various field for data and control and signaling information. These create a subfield to carry various other pieces of information.

Voice Quality Improvements with IS-136 TDMA

One approach to improving the voice quality of IS-136 technology is to improve the performance of voice coders at 8 Kbps. The IS-641 ACELP speech-coding standard provides significant improvement over the initial 8-Kbps VSELP speech coder used with TDMA. Significant transmission delay remains, however, and IS-641 coders in tandem result in degraded performance.

Another approach to improving the voice quality of IS-136 technology is to allocate multiple time-slots to individual users in order to support voice coders at 16 Kbps (two time slots out of three for one frequency channel) or 24 Kbps (all three time slots for a channel). Unfortunately, this has a substantial impact on capacity; it impacts an economical terminal's capability to perform the required signal strength measurements for mobile-assisted-handoff. It also requires terminals to include a duplexer.

A third approach to improving the voice quality of IS-136 technology is to introduce 8-PSK or 16-QAM modulation with efficient channel coding and diversity techniques to provide sufficient robustness in the presence of transmission impairments. This approach could permit the introduction of a higher rate voice coder while maintaining the existing three traffic channels per carrier of IS-136. The 20-ms frame structure could be maintained while reducing transmission delay by removing inter-burst interleaving.

Data Services Currently Available with TDMA

TDMA provides 9.6-Kbps circuit data and fax access based on the IS-130 and IS-135 standards. Enhanced data rates will be important in the future, particularly for Web browsing on the Internet or on corporate Intranets.

The use of data communications always excites the average consumer today. In part, this excitement comes from the lack of high-speed data communications of the past. Enhanced speed access will require additions to the existing standards. The first step is likely to be a multi-slot operation that can support rates up to 28.8 Kbps. A second step is to introduce over-the-air packet access for TDMA in addition to circuit mode access. A further step will be to introduce 8-PSK and/or 16-QAM modes to support data rates up to about 57.6 Kbps. Even higher bit rates may be possible based on transmit and receive diversity. Wireless data access to the Internet is likely to come from stationary but widespread users with laptop computers or *Personal Digital Assistants* (PDA). Under those conditions, simple pre-selection diversity, as proposed for high-quality pedestrian voice service, could significantly improve downlink performance because fading rates will be slow. However, TDMA brings the capability to move both data transmissions and facsimile (fax) traffic to the forefront.

As a matter of fact, the following data communications features become a reality with TDMA:

- Perfect faxes can be delivered in real-time at about two pages per minute.
- File transfers could be up to 115,200 bps with triple-rate channel and compression (depending on data).
- Full-rate data transmission will produce 9.6 Kbps uncompressed or 38.4 Kbps compressed.
- It detects and corrects errors, compresses and encrypts data.
- No special modem or fax machine is required on wireline side.
- The phone looks like a wireline fax/data modem.
- Future Windows unimodem will also contain built-in support for TDMA data communications.
- It is compatible with existing software.

Automatic and Dynamic Channel Assignment for IS-136

Most TDMA systems are based on conventional *Fixed Channel Assignment* (FCA). The exceptions to this are emerging wireless office systems, personal base stations, and microcellular systems. Much work has been done to examine the use of *automatic channel assignment* (ACA) and *dynamic channel assignment* (DCA) for AMPS, TDMA, and other cellular systems. DCA has been considered for mobile systems from the early days of research on cellular telephony. The motivations for DCA include spectrum efficiency, dealing with hot spots, and reducing manual frequency planning. Using per-carrier amplifiers and cavity combiners at base stations has been a major impediment to the introduction of DCA for cellular systems. Other impediments include difficulty in implementing DCA with very large systems, performance under heavy load conditions, and challenges with evolving from FCA to DCA. However, several factors suggest that DCA will become a core capability for TDMA systems:

- Emergence of hierarchical cellular systems
- Development and deployment of automatically tuned cavity combiners and multicarrier amplifiers
- Experience learned from early adaptors of DCA techniques

TDMA uses channels that are 30 kHz wide. This makes TDMA particularly well suited for dynamic channel allocation for microcells, personal base stations, and wireless office systems within a macrocellular system that either uses FCA or DCA. With 30-kHz channels, microcells and indoor systems that are uncoordinated or loosely coordinated with a macrocellular system can assign channels in small increments. This is particularly useful for the coexistence of a large number of overlapping or nearby systems that are uncoordinated except for frequency assignment arrangements.

Macrocells typically transmit with much higher power than microcells, so microcells can readily measure the signals of nearby macrocells for ACA; however, macrocells cannot easily measure the signals of nearby microcells, so macrocells must depend on rules or databases instead of measurements to avoid interfering with microcells upon making new channel assignments.

Microcellular Evolution for IS-136

Microcells today share much of their architecture with macrocells. In the future, TDMA microcells will be more highly optimized for small cell operation based on several trends including the following factors:

- Centralization of functionality
- Mass miniaturization of the components and antennae
- Significant power reductions
- Interconnectivity between and among systems
- Line powering for picocells
- Automatic channel assignment

The optimization of microcells for small-cell operation will make TDMA more cost-effective and flexible for deployment in densely populated areas and indoor environments where a private system can be deployed. T-1 service over ordinary unloaded copper loops at distances of 12 kilofeet is quickly becoming reasonably priced. This drop in cost provides a low-end backhaul service for microcells in the future. Using the 64-Kbps PCM channel for the microcell provides a robust alternative to many of today's existing impediments.

Future High-Speed Packet Data Wireless Access Using EDGE

Web browsing and information service access, which has caused the recent explosion in Internet usage, is highly asymmetrical in transmission requirements. Only the transmission path to the subscriber needs to be high speed for many applications. Many other services provided over the Internet can also be provided with low- to moderate-bit rate uplinks. However, large file transfer is an example of an application that benefits from symmetrical high-speed data systems.

The *Universal Wireless Communications Consortium* (UWCC) has adopted *Enhanced Data rates for GSM Evolution* (EDGE) based on enhancing GSM packet data technology with adaptive modulation to support packet data access at speeds of up to about 384 Kbps. This system will adapt between GMSK and 8-PSK modulation with up to eight different channel-coding rates to support packet data communications at high-speeds on low interference/noise channels and at lower speeds on channels with heavy interference/noise. EDGE can be used with conventional 4/12 reuse as is used for GSM. However, the use of 1/3 reuse is also proposed to permit the initial roll out with only 2×1 MHz of spectrum. One challenge for 1/3 reuse is the operation of control channels that require robust operation. Synchronizing the frame structures of base station transceivers throughout a network and using time-reuse to achieve adequate protection for control channels can accomplish this. GPRS networking is proposed to support EDGE-based wireless packet data access.

UWC-136 was designed to provide a 136-based radio transmission technology candidate that meets ITU-R's requirements for IMT-2000. UWC-136 maintains the TDMA community's philosophy of evolution from first- to third-generation systems while addressing the specific desires and goals of the TDMA community for a third-generation system. TR45.3 believe that UWC-136 is an attractive and powerful evolutionary step for 136. The technology presented provides for future IMTS services to existing operators, as well as providing new operators with competitive features, services, and technology. Additionally, the technology provides these same features and services in other bands around the world where regulatory approval has been granted to offer such services, primarily in the 450-, 800-, and 1,900-MHz bands.

UWC-136 meets IMT-2000 objectives via modulation enhancement to the existing 30-kHz 136 channel (136+) and defines complementary wider-band TDMA carriers (136 HS) to address those services that are not possible on the 30-kHz carriers. Together, the existing deployed system (136), the enhanced 30-kHz 136 carrier (136+), and the complementary wider-band TDMA carriers to address the High-Speed (136 HS) aspect of IMT-2000 are referred to as UWC-136. It is important to realize that the technologies submitted as UWC-136 for IMT-2000 are band independent and may be deployed by existing or new operators. TR45.3 does not subscribe to the common belief that IMT-2000 is limited to a specific RF band; rather it believes that it is a service, independent of the bands in which it is deployed.

Additional Considerations for 136 HS

The 136 HS air interface was developed to satisfy the requirements for an IMT-2000 radio transmission technology, with additional requirements for the consideration of a commercially effective evolution and deployment in current 136 networks. Such considerations include flexible spectrum allocation, spectrum efficiency, compatibility with 136 and 136+, and support

Figure 3-10
Evolution to IMT-
2000 in IS-136

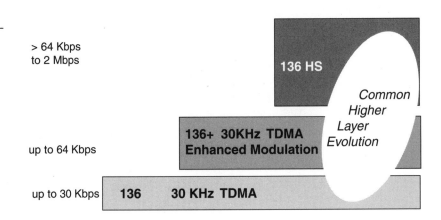

BEARER RATE

of macrocellular performance at higher mobile speeds. Several of these requirements are above and beyond the ITU requirements for IMT-2000, which are crucial to 136 operators. The evolution of the IS-136 standards to the higher bandwidth applications is shown in Figure 3-10.

Extended-TDMA

Standards bodies and manufacturers all came up with variations of the use of the frequency spectrum. TDMA has been discussed and pretty well settled upon in the past. Nevertheless, manufacturers have been developing an *Extended-TDMA* (ETDMA), which will enable efficient use of the system. TDMA will derive a three- to five-fold increase in spectrum use, whereas ETDMA could produce 10 to 15-fold increases. The concept is to use a *Digital Speech Interpolation* (DSI) technique that reallocates the quiet times in normal speech, thereby assigning more conversations to fewer channels, gaining up to 15 times over an analog channel. This is a form of statistical time division multiplexing. When a device has something to send, it places a bit in a buffer. As the sampling device sees the data in the buffer, it allocates a data channel to that device. When a device has nothing to send, nothing is placed into the buffer. Then, the sampling passes over a device with an empty buffer. Time slots are dynamically allocated based on need rather than on fixed time-slot architecture. However, standards bodies and manufacturers all came up with variations of the use of

Figure 3-11
ETDMA uses a form of statistical time division multiplexing.

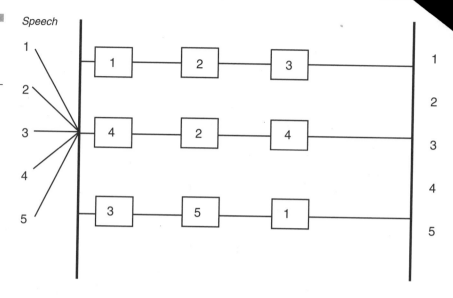

the frequency spectrum. Figure 3-11 provides an example of the ETDMA technique.

Code Division Multiple Access (CDMA)

CDMA is a radical shift from the original FDMA and TDMA wireless techniques. This system has been gaining widespread acceptance across the world in the cellular industry. The cellular providers see CDMA as an upgrade opportunity for their capacity and quality. CDMA is a form of spread spectrum, a family of digital communications techniques. The core principle behind CDMA is the use of the noise-floor to carry radio signals. As its name implies, bandwidth greater and wider than normal constrained FDMA and TDMA channels is used. Point-to-point communications is effective on the bandwidth that uses the noise waves to carry the signal spread across a significantly wider radio carrier.

Spread spectrum, which was employed back in the 1920s, has evolved from military security applications. It uses a technique of organizing the radio frequency energy over a range of frequencies, rather than a modulation technique. The system uses frequency hopping with time division multiplexing. At one minute the transmitter is operating at one frequency, at the next instant it is on another. The receiver is synchronized to switch frequencies in the same pattern. This is effective in preventing detection (interception) and jamming; thus, additional security is derived. These techniques should produce increased capacities of 10- to 20-fold over existing analog systems.

Originally conceived for commercial application in the 1940s, it was an additional 40 years before this technique became commercially feasible. The main factors holding back the use of CDMA were cost and complexity of operation. Today, the use of low-cost, high-density digital *integrated circuits* (ICs) that reduce the size and weight of the radio equipment makes the use of CDMA far more feasible. Another area is an educational one, whereby the carriers needed to understand that the use of optimal communications requires that the station equipment must regulate the power to the lowest possible levels to achieve the maximum performance. This, of course, flies in the face of the normal operations that the carriers received their training.

In 1989, spread spectrum CDMA was commercially introduced as the solution to the bandwidth demands of the industry. By using a spread spectrum frequency-hopping technique, developers announced that they could achieve the desired frequency reuse patterns everyone was looking for. This was already being used in the satellite transmission systems and some of the *Specialized Mobile Radio* (SMR) organizations.

Spread spectrum can use one of two different techniques: *frequency hopping* (FH) or *direct sequence* (DS). In both cases, the synchronization between the transmitter and receiver is crucial. Both forms use a pseudo-random carrier; they just do it in different ways.

■ Frequency hopping is usually accomplished by the rapid switching of fast-settling frequency synthesizers in a pseudo-random pattern. This is not usually implemented in the commercial versions of CDMA.

■ Direct sequence is used by commercially available CDMA. It is accomplished as a multiple of the more conventional waveform by using a pseudo-noise binary sequence at the transmitter. Noise and interference across the waveform are uncorrelated with the pseudo-noise sequence, and thus become like noise. This increases the bandwidth when they reach the detectors. The *signal-to-noise ratio* (SNR) can be enhanced by using filters that reject the interference.

The CDMA Cellular Standard

With CDMA, unique digital codes, rather than separate radio frequencies or channels, are used to differentiate subscribers. The codes are shared by both the mobile station (cellular phone) and the base station, and are called *pseudo-random code sequences*. All users share the same range of radio spectrum. For cellular telephony, CDMA is a digital multiple access technique specified by the *Telecommunications Industry Association* (TIA) as IS-95. In March 1992, the TIA established the TR-45.5 subcommittee with the charter of developing a spread-spectrum digital cellular standard. In July of 1993, the TIA gave its approval of the CDMA IS-95 standard. IS-95 systems divide the radio spectrum into carriers that are 1,250 kHz (1.25 MHz) wide. One of the unique aspects of CDMA is that although the number of phone calls that a carrier can handle is certainly limited, it is not a fixed number. Rather, the capacity of the system will be dependent on a number of different factors.

CDMA changes the nature of the subscriber equipment from a predominantly analog device to a digital device. Older radio systems separate the channels or stations on the RF with filters in the frequency domain. CDMA receivers do not eliminate analog processing in its entirety, but they separate the communications channels by a pseudo-random modulation technique that is applied to the digital domain, not based on frequencies. In fact, multiple users will occupy the same frequency band simultaneously.

To achieve an acceptance of CDMA, the industry must understand that it changes the way we do business, as outlined in Table 3-1.

Spread Spectrum Goals

Much of the action in the wireless arena includes the use of the frequency spectrum to its fullest while preserving its efficiency. The primary goal of spread spectrum systems is the substantial increase in bandwidth of an information-bearing signal, greater than required for basic communications. This increased bandwidth, though not used for carrying the signal, can mitigate possible problems in the airwaves, such as interference or inadvertent sharing of the same channels. The cooperative use of the spectrum is an innovation that was not commercially available in the past.

Table 3-1

Reasons for Using CDMA in Cellular

CDMA alters the way we communicate by:
Improving the telephone capacity of cellular operators
Improving quality of the voice communications and eliminating audible impairments of multipath fade
Reducing the incidence of dropped calls especially during handoff
Providing reliability for data communications, that is, fax and Internet traffic
Reducing the number of sites to support a specific volume of traffic
Simplifying the site selection process
Reducing the average transmitter power output requirements
Reducing or eliminating interference with other electronic devices in the area
Limiting potential health risks
Reducing the operating costs because fewer cell sites are needed

Additional goals of CDMA and in particular, multiple access communications, include the following:

- Near-wireline quality
- As close to universal coverage as possible
- Low equipment costs
- Minimum number of sites

These goals were derived from Table 3-1 and help to keep focused on the overall responsibilities and uses of the frequency spectrum. Regulators around the world have set aside limited amounts of bandwidth to satisfy these services, so that the efficiency is kept high. The limited frequency spectrum allocated preserves upon the goal of using spectral efficiency, which is usually measured with one of the traffic engineering calculations (Erlang or Poisson) per unit in operation in a specified geography and in terms of per MHz. For example, cellular operators use a 25-MHz split between the two directions of communications: 12.5 MHz of transmit and 12.5 MHz of receive spectrum. As technology enhancements occur, practical ways of expanding the amount of coverage become a reality.

Spread Spectrum Services

As the use of radio frequency spectrum continued to put pressure on this limited resource, the manufacturers of systems and regulators were searching for some way to share spectrum among multiple users. Furthermore, sharing is compounded by the need to secure information while on airwaves. These pressures have led to the use of spread spectrum radio. The spreading portion of these systems using a chip set coded for your specific transmitter to receiver system uses multiple frequencies (called hopping) as one method; or another technique of creating a coded chip set is used. Either way, both of these services are designed to spread as much energy over a broader range of frequencies to enable less airtime on a specific bandwidth and to ensure the integrity of the information being sent. The technique for the spread spectrum service is called Code Division Multiple Access (CDMA). Many of the PCS carriers have chosen CDMA as their coding of choice. The concept of CDMA is shown in Figure 3-12.

Because spread spectrum has been introduced at the commercial level, the FCC allocated spectrum in the 1.9-GHz range. This group of frequencies is heavily used by microwave users operating in the 2-GHz range. The FCC felt that the spread spectrum would not seriously impair the microwave users because it will appear only as random noise to any other fixed system operating in the same frequency ranges. However, as more users of spread spectrum are inserted into a specific frequency range, the possibility of congestion and interference exists. If you attempt to use this range of frequencies and congestion builds, then the decision could be a wrong one.

Figure 3-12
CDMA spread
spectrum services

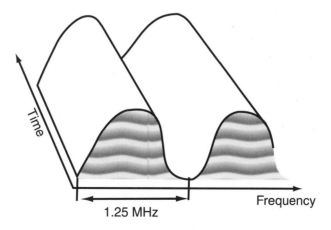

Although frequency hopping has its benefits, distance, utilization, and power constraints may overshadow these benefits.

Security

Increased privacy is inherent in CDMA technology. CDMA phone calls will be secure from the casual eavesdropper because unlike an analog conversation, a simple radio receiver will not be able to pick individual digital conversations out of the overall RF radiation in a frequency band.

Synchronization

In the final stages of the encoding of the radio link from the base station to the mobile, CDMA adds a special pseudo-random code to the signal that repeats itself after a finite amount of time. Base stations in the system distinguish themselves from each other by transmitting different portions of the code at a given time. In other words, the base stations transmit time-offset versions of the same pseudo-random code. In order to assure that the time offsets used remain unique from each other, CDMA stations must remain synchronized to a common time reference.

The primary source of the very precise synchronization signals required by CDMA systems is the *Global Positioning System* (GPS). GPS is a radio navigation system based on a constellation of orbiting satellites. Because the GPS system covers the entire surface of the earth, it provides a readily available method for determining position and time to as many receivers as are required.

Balancing the Systems

CDMA cell coverage is dependent upon the way the system is designed. In fact, three primary system characteristics—coverage, quality, and capacity —must be balanced off of each other to arrive at the desired level of system performance.

In a CDMA system, these three characteristics are tightly interrelated. Even higher capacity might be achieved through some degree of degradation in coverage and/or quality. Because these parameters are all intertwined, operators cannot have the best of all worlds: three times wider coverage, 40 times capacity, and CD-quality sound. For example, the

13-Kbps vocoder provides better sound quality, but reduces system capacity as compared to an 8-Kbps vocoder.

Operators will have the opportunity to balance these parameters to best serve a particular area. The best balance point may change from cell site to cell site. Sites in dense downtown areas may trade off coverage for increased capacity. Conversely, at the outer edges of a system, capacity could be sacrificed for coverage area.

Common Air Interfaces

Two primary air interface standards are in use today:

- Cellular (824–894 MHz) uses the TIA/EIA/IS-95A
- PCS (1,850–1,990 MHz) uses the ANSI J-STD-008

These two standards are similar in the features that they offer, with the exception of the frequency plan, mobile identities, and message fields. The standards provide some stability in the operation of the systems, but may change over time. However, looking at the forward and reverse CDMA channel can shed some added light on what we can expect from the use of CDMA.

The Forward Channel

The forward CDMA channel is the cell site to mobile direction for communications. It carries traffic, a pilot signal, and any overhead information required by the system. The pilot is a spread, but otherwise unmodulated DSSS signal. The pilot and overhead channels establish and maintain the system timing and the station identity. The pilot is also used in the *mobile-assisted handoff* (MAHO) process as a signal strength indicator.

Transmission Speeds and Rates   IS-95A uses a forward link that supports a speed of 9,600 bps in the data-bearing channels. The forward error correction code rate is 1/2 and the pseudonoise rate is 1.2288 MHz (which is $128 \times 9,600$ bps). Table 3-2 represents the forward link channel parameters, with a rate set 1 from IS-95A.

The J-STD-008 shown in Table 3-3 supports an additional set of parameters with a maximum transmission rate of 14,400 bps. This is called rate

Table 3-2

Forward Link
Channel Parame-
ters IS-95A: Rate
Set 1

Channel	Sync	Paging		Traffic				Rate
Data rate	1,200	4,800	9,600	1,200	2,400	4,800	9,600	bps
Code repetition	2	2	1	8	4	2	1	
Modulation symbol rate	4,800	19,200	19,200	19,200	19,200	19,200	19,200	Sps
PN chips/ modulation symbol	256	64	64	64	64	64	64	
PN chips/ bit	1,024	256	128	1,024	512	256	128	

Table 3-3

Forward Link
Channel Parame-
ters J-STD-008: Rate
Set 2

Channel		Traffic			Rate
Data rate	1,800	3,600	7,200	14,400	bps
Code repetition	8	4	2	1	
Modulation symbol rate	19,200	19,200	19,200	19,200	Sps
PN chips/ modulation symbol	64	64	64	64	
PN chips/ bit	682.67	341.33	170.67	85.33	

set 2. This set uses a FEC code rate of 3/4 created by puncturing the previous code.

The forward link is comprised of up to 64 logical channels (code channels). The channels are independent in that they carry different data streams, even possibly at different rates, and they are adjustable in amplitude.

Overhead Channels

Three different types of overhead channels exist in the forward link. These include the pilot, sync, and paging channels. The pilot is a requirement for every station.

- **Pilot channel**   This is always code channel 0. It operates a demodulation reference for the mobile stations and a handoff level measurement reference. Therefore, it is a required reference in all stations. The pilot does not carry information.

- **Sync channel**   This carries a repeating message that identifies the individual station and the absolute phase of the pilot sequence. The data rate of the sync channel is 1,200 bps. This mobile finds the framing boundary of the sync channel and times to it simply. It carries a single repeating message that conveys timing and system configuration to the mobile station. The mobile station can therefore derive accurate system time by synchronizing to the short code.

- **Paging channel**   This is used to communicate with the mobile stations when they are not assigned to a traffic channel. Its purpose is to notify the mobile of incoming calls and carries responses to the mobile access whereby an assignment of a traffic channel may occur. Paging channels operate at 4,800 or 9,600 bps.

- **Traffic channels**   These are dynamically assigned channels in response to a mobile access. The paging channel is used to notify the mobile of which code channel it should use to receive. The traffic channel carries its data in a 20-ms frame.

The Reverse Channel

The reverse channel is the mobile-to-cell site communication channel. It carries traffic and signaling information. A reverse channel is only active during calls associated with a specific mobile station or when access channel signaling takes place to the base station.

Transmission Speeds and Rates   IS-95A uses a reverse link that supports a speed of 9,600 bps in the access and traffic channels. The forward error correction code rate is 1/3, the code symbol rate is 28,800 symbols per second after six code symbols per modulation symbol are present, and the pseudo-noise rate is 1.2288 MHz. Table 3-4 is a representation of the reverse link channel parameters, with a rate set 1 from IS-95A.

The J-STD-008 shown in Table 3-5 supports an additional set of parameters with a maximum transmission rate of 14,400 bps. This is called rate set 2. This set uses a FEC code rate of 1/2 in place of the rate 1/3 code of rate set 1.

Table 3-4

Reverse Channel Parameters IS-95A: Rate Set 1

Channel	Access	Traffic				Rate
Data rate	4,800	1,200	2,400	4,800	9,600	bps
Code rate	1/3	1/3	1/3	1/3	1/3	
Symbol rate before repetition	14,400	3,600	7,200	14,400	28,800	Sps
Symbol repetition	2	8	4	2	1	
Symbol rate after repetition	28,2800	28,800	28,800	28,800	28,800	Sps
Transmit duty cycle	1	1/8	1/4	1/2	1	
Code symbols/ modulation symbols	6	6	6	6	6	
PN chips/ modulation symbol	256	256	256	256	256	
PN chips transmitted/bit	256	128	128	128	128	

Table 3-5

Reverse Link Channel Parameters J-STD-008: Rate Set 2

Channel	Traffic				Rate
Data rate	1,800	3,600	7,200	14,400	bps
Code rate	1/2	1/2	1/2	1/2	
Symbol rate before repetition	3,600	7,200	14,400	28,800	Sps
Symbol repetition	8	4	2	1	
Symbol rate after repetition	28,800	28,800	28,800	28,800	Sps
Transmit duty cycle	1/8	1/4	1/2	1	
Code symbols/ modulation symbol	6	6	6	6	
PN chips/256 modulation symbol	256	256	256		
PN chips transmitted/ bit	256/3	256/3	256/3	256/3	

Channelization The reverse CDMA channel consists of $2^{42} - 1$ logical channels. One of these channels is permanently and uniquely associated with each mobile station. The mobile uses the logical channel whenever it passes traffic. The channel does not change upon a handoff. Other logical channels are used for access with the base station.

Traffic Channel

Traffic channels are the reverse CDMA channels and are mobile unique. The traffic channel always carries data in a 20-ms frame. Frames at the higher rates of rate set 1 and all those in rate set 2 use CRC codes to help assess the frame quality in the receiver.

Cellular Communications

The world, as we now know it, is changing at a terrific rate due to technology advances. The year 1984 marked a major milestone for the telecom industry in the twentieth century. The AT&T monopoly was broken up and it was the year we were first introduced to wireless service.

Much has happened in the cellular industry around the world since its original introduction in 1984. In 1984, cellular communications became the hot button in the industry. At the time, all systems used analog radio transmissions. Why did they choose analog? It was all that the providers had available at the time, and digital was seen as a future innovation. Digital has its detractors as well. Roaming may be more difficult using a digital-based phone than an analog. Because there is no single accepted industry standard in digital technology today and the technologies are incompatible, roaming (using another wireless operator's network while traveling) may be difficult. Analog has better coverage and greater service availability than digital. The initial cost for analog is usually cheaper than for digital or at least the proponents will have you believe this.

As you've probably noticed, price wars abound, involving all wireless operators. In fact, in some areas, wireless operators are offering pricing plans that rival even landline services. The average monthly bill of a wireless customer has been slashed in half in the past decade, just shy of $40 from $95, and the cost per minute has dropped to an average of less than $.10 from $.60 cents when the cellular services were commercially introduced. PCS operators may offer lower prices because some markets have more than five operators, causing the competition to create price wars.

In the end, the customer must choose what will work best for his or her needs. Nevertheless, it is important to remember that the wireless company has just as big of an incentive to retain you as you have in choosing them. It is still necessary to ascertain what future plans a provider has for wireless communication. If he or she will provide data services and if voice quality and security is of utmost importance, then it may be best to invest in a digital phone service plan. If roaming accommodations is a top priority, then using a cellular provider may be the answer for now. To cover all the bases, make sure that you have a dual-mode phone capability so that although digital service may be preferred, when roaming or when out of digital territory, which may be pretty frequent, analog service is available.

Analog networks proliferated across the country creating ubiquitous coverage. Many reasons were used to justify the cellular networks. These included

■ Very limited service areas—you could not get service where you needed it.

■ Poor transmission haunted the operators because of the nature of the radio systems.

■ Excessive call setup delays.

■ Heavy demand building at a rapid pace.

■ Limited channel availability from older mobile networks.

Analog cellular radio systems used *frequency division multiple access* (FDMA), which is an analog technique designed to support multiple users in an area with a limited number of frequencies. Analog radio systems use analog input, such as voice communications. Because these systems were designed around voice applications, no one had any thought for the future transmission of data, fax, or packet data. Moreover, high-speed, broadband services were not even on the radar screen of the manufacturers and carriers.

Initially, no one was sure what the acceptance rate would be in 2001, approximately 200 million people were cellular users in North America. On average, over 13 million users are signing up for some form of wireless communications per month throughout the world. Some expect that in 2008 North America will have over 1 billion users and the rest of the world will have 1.8 billion users, as seen in Figure 4-1. Furthermore, it is estimated that four of five new telephones sold today are wireless telephones. What was once a business tool has become an indispensable device to the masses. Consequently, we can ignore the acceptance factor as a non-issue. The new problem is not one of acceptance, but retaining users. The churn ratio has always been as high as 15 to 20 percent.

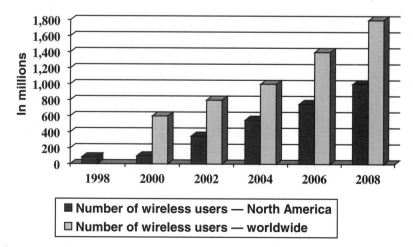

Figure 4-1
Predictions of wireless users by 2008

Yet, other people predict that the numbers will look differently, with several different research houses pushing the envelope. Figure 4-2 presents a different projection that shows 1 billion users in operation in 2003.

In the U.S. alone, the figures are rapidly growing where the number of wireless telephones are expected to reach 200 million new sets in 2002 and rapidly grow from there, as seen in Figure 4-3.

The answer to the retention problem was in packaging the service with the handset. The driving factors for the acceptance and reduced churn rates are as follows:

- Service plans with a usage fee of $.10 /minute
- Greater geographic coverage
- Elimination of roaming charges
- Elimination of toll charges

This comes from that old adage packaged by the Gillette Company years ago: "give them the razors, they will buy the blades." In the cellular world, if you give people the wide area and free long distance, they will buy the service, as we have seen.

Figure 4-2
Growth predictions of wireless and PC users

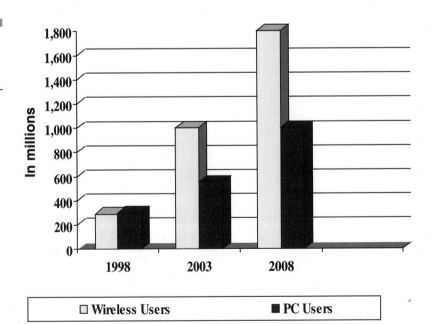

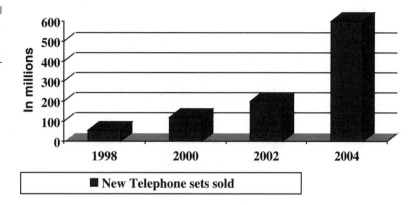

Figure 4-3
New telephones sold
in U.S. alone (est.)

The cellular industry is still primarily running on an analog backbone. Estimates are that 100 percent of the U.S. has analog coverage, whereas the digital counterpart to the cellular networks only covers between 60 to 100 percent depending on the technologies used. In Canada, the coverage is also mainly analog because little opportunity exists for the upgrades to digital technologies due to the population in the areas where the systems are installed. However, the estimates are that only 35 to 40 percent of the U.S. population has a mobile phone. Contrasting this number to the number of households that have access to the Internet, the answer is an astounding 54 percent. The wireless carriers generate more than $44 billion revenue annually. Only $4 billion of that revenue comes from roaming fees. The wireless providers in the U.S. employ approximately 150,000 people. Nearly 50 percent of the companies in the U.S. offer their products online.

Dual-mode telephones have become the salvation to cellular providers because they would not be able to sustain their customer base without this offering. Still some users will not part with their existing services, albeit analog cellular services. In Figure 4-4, we see the percentage of analog and digital subscribers for the U.S. as of the end of 2000. In 1999, digital penetration shifted and the number of digital subscribers exploded.

Why Digital?

Many nationwide wireless service providers have a digital network in place or are at least in the process of converting their cellular network to a digital

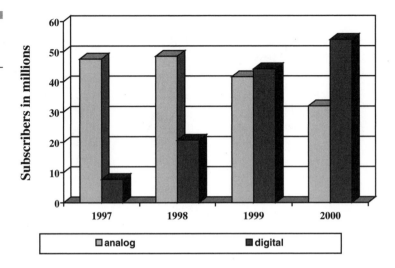

network. Despite the arguments to the contrary, digital presents many advantages over cellular. The main benefits of digital include better quality of service, higher security, and the capability to support next-generation services. Operators are looking to digital technology to help enable the gee-whiz services, including the wireless Internet.

Digital is known to increase the efficiency in the network, meaning an operator can fit more information into each transmission; this is why so many operators are now converting their systems to digital. Wireless operators can get more bang for their buck with digital networking.

Digital offers a better quality of sound. Proponents of digital also claim that because digital scrambles up the signals into bursts, it is more secure than analog and can help thwart *cloning*, the act of grabbing phone account information over the air in order to copy and resell that information for piracy purposes. Digital handsets also support longer battery life than analog.

Coverage Areas

The cellular operators built out their networks to provide coverage in certain geographically bounded areas. This poses a dilemma for the providers:

■ The carriers need (more) users to generate higher revenues to pay off their investment.

■ They must continue the evolution from analog to digital systems, enabling more efficient bandwidth use.

■ Security and protection against theft is putting pressure on the carriers and users alike.

Analog Cellular Systems

Analog systems do nothing for these needs. Using amplitude or frequency modulation techniques to transmit voice on the radio signal uses all of the available bandwidth. This means the cellular carriers can support a single call today on a single frequency. The limitations of the systems include limited channel availability.

The analog system was designed for quick communications while on the road. Because this service would meet the needs of users on the go, the thoughts of heavy penetration were only minimally addressed. However, as the major *Metropolitan Service Areas* (MSA) began expanding, the carriers realized that the analog systems were going to be too limiting. With only a single user on a frequency, congestion in the MSA became a tremendous problem. A cellular channel uses 30 kHz of bandwidth for a single telephone call!

Cellular was designed to overcome the limitations of the conventional mobile telephone. Areas of coverage are divided into honeycomb-type cells, hexagonal design, of smaller sizes shown in Figure 4-5. The cells overlap each other at the outer boundaries. Frequencies can be divided into bands or cells to prevent interference and jamming of the neighboring cell's frequencies. The cellular system uses much less power output for transmitting. The vehicular transmitter uses 3 watts of power, whereas the handheld sets use only 3/10 watts. Frequencies can be reused much more often and closer to each other. The average cell design is approximately 3 to 5 miles across. The more users subscribe to a network, the closer the transmitters are placed to each other. In rural areas, the cells are much further apart.

Cell sites may be separated by 3 to 5 miles for normal operation, but as more users complain of no service due to congestion, cell splitting occurs. A cell can be subdivided into smaller cells, reallocating frequencies for

Figure 4-5
The cell patterns

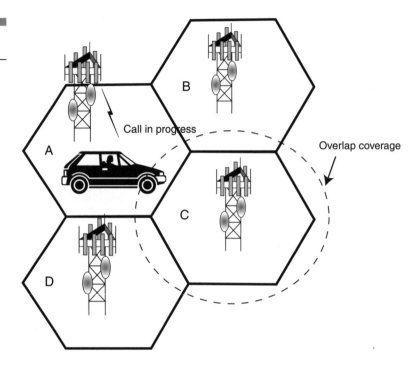

continued use. The smaller the cell, the more equipment and other components are necessary. This placed an added financial burden on the carriers as they attempted to match customer need with return on investment.

Log On

When the vehicle telephone powers on, it immediately logs onto the network. First, the telephone set sends a message to the *Mobile Telephone Switching Office* (MTSO). The MTSO is the equivalent of a Class 5 Central Office (CO). It provides all the line and trunk interface capabilities in much the same way as the CO does.

The information sent to the MTSO includes the electronic serial number and telephone number from the handset. These two pieces of information combined will identify the individual device.

The telephone set will use an information channel to transmit the information. Several channels are set aside specifically for the purposes of log-on capabilities, as shown in Figure 4-6.

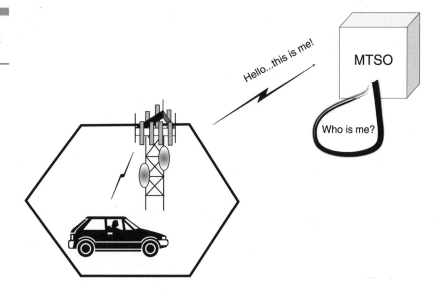

Figure 4-6
Log-on process uses
specific channels

Monitoring Control Channels

Once a telephone set has logged on, it will scan the 21 channels set aside as control channels. Upon scanning, the telephone will lock in on the channel that it is receiving the strongest. It will then go into monitoring mode. Although the set has nothing to send, it will continue to listen to the monitored channel in the event that the MTSO has an incoming call for it. The telephone user has actually done nothing. Upon powering up, the set immediately logged onto the network, identified itself, and went into the monitoring mode.

Failing Signal

One can assume in vehicular communications that the vehicle will be in motion. As the vehicle moves from cell to cell, it will move out of range from the first site, but come into range of the second site. The received signal on the monitoring channel will begin to fail (become too weak to hear), as shown in Figure 4-7. Immediately, the telephone will re-scan all the monitoring channels and select a new one. Once the vehicle finds a new channel, it will continue to monitor that channel until such time that it rolls

Figure 4-7
The failing signal
procedure

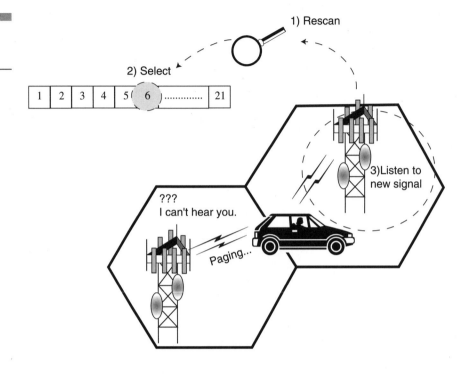

out of range again. This concept of rolling from site to site enables the vehicle to be in constant touch with the MTSO so long as the set is on.

Setup of a Call

When the user wishes to place a call, the steps are straightforward. The process is very similar to making a wired call:

1. Pick up the handset and dial the digits.
2. After entering the digits, press the send key.
3. The information is a dialogue between the MTSO and the handset.
4. The MTSO receives the information and begins the call setup to a trunk connection.
5. The MTSO scans the available channels in the cell and selects one.
6. The MTSO sends a message to the handset telling it which channel to use.

7. The handset then tunes its frequency to the assigned channel.

8. The MTSO connects this channel to the trunk used to set up the call.

9. The call is connected and the user has a conversational path in both directions.

Setup an Incoming Call

When a call is coming in from the network, the process is again similar to the wireline network.

1. The mobile office receives signaling information from the network that a call is coming in.

2. The mobile office must first find the set, so it sends out a page through its network.

3. The page is sent out over the control channels.

4. Upon hearing the page, the set will respond.

5. The mobile office hears the response and assigns a channel.

6. The mobile office sends a message telling the set that it has a call and to use channel X.

7. The set immediately tunes to the channel it was assigned for the incoming call.

8. The phone rings and the user answers.

Handoff

While the user is on the telephone, several things might happen. The first is the vehicle could move away from the center of the cell site. Therefore, the base station must play an active role in the process of handling the calls.

1. As the user gets closer to the boundary, the signal will become weaker.

2. The base station will recognize the loss of signal strength and send a message to the mobile office.

3. The mobile office will go into recover mode.

4. The MTSO must determine what cell will be receiving the user.

5. The MTSO sends a message to all base stations advising them to conduct a quality of signal measurement on the channel in question.

6. Each base station determines the quality of the received signal.

7. They will advise the MTSO if the signal is strong or weak.

8. The MTSO decides which base station will host the call.

Setting up the Handoff

Once the MTSO has determined which base station will be the new host for the call, it will select a channel and direct the new base station to set up a talk path for the call. This is all done in the background. An idle channel is set up in parallel between the base station and MTSO.

The Handoff Occurs

1. The original base station is still serving the call.

2. The new base station will host the caller.

3. The parallel channel has been set up.

4. MTSO has notified the cells to set the parallel channel in motion.

5. The MTSO sends a directive to the telephone to retune its frequency to the new one reserved for it.

6. The telephone set moves from one frequency to the new one.

7. The call is handed off from one cell to another.

8. The caller continues to converse and never knows what happened.

This procedure is shown in Figure 4-8.

Completion of the Handoff

Once the telephone has moved from one base station to the other and one channel to another, the handoff is complete. However, the original channel is now idle, but in parallel to the original call. Therefore, the base station notifies the MTSO that the channel is now idle. The MTSO is always in control of the call. It manages the channels and the handoff mechanisms. The MTSO commands the base station to set the channel to idle and makes it available for the next call.

Figure 4-8
The handoff process

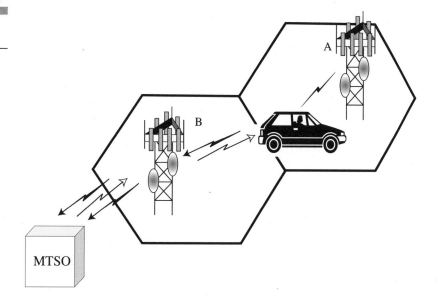

The Cell Site (Base Station)

The preceding discussion centered on the process of the call and referred to the base station quite a bit. The cell is comprised of 3- to 5-mile radius. The base station is comprised of all the transmission and reception equipment between the base station and MTSO and the base station to the telephone. The cell has a tower with multiple antennae mounted on the top. Each cell has enough radio equipment to service approximately 45 calls simultaneously, as well as monitor all the channels in each of the adjacent cells to it. This is shown in Figure 4-9. The equipment varies with the manufacturer and the operator, but an operator will typically have 35 to 70 cells in a major location.

The Mobile Telephone Switching Office (MTSO)

The MTSO is a Class 5 Central Office equivalent. It provides the trunks and signaling interfaces to the wireline carriers. It has a full line switching component and the necessary logic to manage thousands of calls

Figure 4-9
The cell call layout

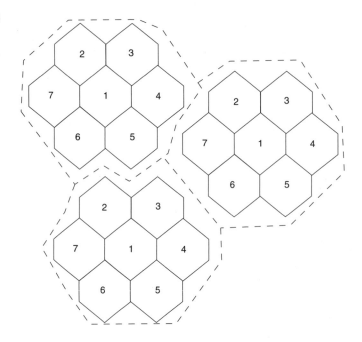

simultaneously. Like the CO infrastructure, the MTSO uses digital trunks between the MTSO and the wireline carriers (ILEC, CLEC, or IEC) either on copper, fiber, or microwave radio systems.

At the MTSO, a separate trunk/line interface exists between the MTSO and the base station. This is the line side of the switch and is used for the controlling call setup. Normally, the MTSO connects to the base station via a T1 operating at 32 Kbps ADPCM. This T1 will be on copper or microwave. A MTSO is a major investment, ranging from $2 to $6 million depending on the size and the area being served.

Frequency Reuse Plans and Cell Patterns

Frequency reuse is what started the cellular movement. Planning permits the efficient allocation of limited radio frequency spectrum for systems using frequency-based channels (AMPS, DAMPS, and GSM). Frequency reuse permits increases in capacity and avoids interference between sites sharing the frequency sets. Frequency plans exist that specify the division of channels among three, four, seven, and twelve cells. They define the organization of available channels into groups, which maximizes service and minimizes interference.

As a mobile unit moves through the network, it is assigned a frequency during transit through each cell. Because each cell pattern has one low-power transmitter, air interface signals are limited to the parameters of each cell. Air interface signals from nonadjacent cells do not interfere with each other. Therefore, a group of nonadjacent cells can reuse the same frequencies.

CDMA systems do not require frequency management plans because every cell operates on the same frequency. Site resources are differentiated by their PN offset (phase offset of the pseudo-random noise reference). Mobile channels are identified by a code used to spread across the baseband signal, and each can be reused in any cell. Using an N=7 frequency reuse pattern, all available channels are assigned to their appropriate cells. It is unnecessary to deploy all radios at once, but their use has been planned ahead of time to minimize interference in the future. Figure 4-10 illustrates this pattern.

Figure 4-10
Seven-cell pattern

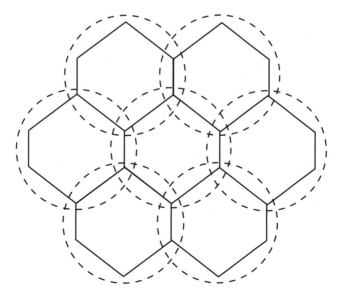

Overlapping Coverage

Each cell has its own radio equipment with an overlap into adjoining cells. This allows for the monitoring of the adjacent cells to ensure complete coverage. The cells can sense the signal strength of the mobile and handheld units in their own areas and in the overlap areas of each adjoining cell. This is what makes the handoff and coverage areas work together.

Cell Site Configurations

The mode of operation of a cell site is determined by the type of antenna used to support the air interface between the cell site and the mobile phones. When an omnidirectional antenna is used, the site serves a single 360-degree area around itself. Figure 4-11 illustrates the omnidirectional antenna.

The cell sites using the omnidirectional antenna are supported by a single antenna for both send and receive operations. These devices cover the full 360-degree site independently. One transmit antenna is used for each radio frame at the site (one frequency group per radio frame). Two receive antennas distribute the receive signal to every radio, providing diversity reception for every receiver at the site.

Figure 4-11
The omnidirectional
antenna

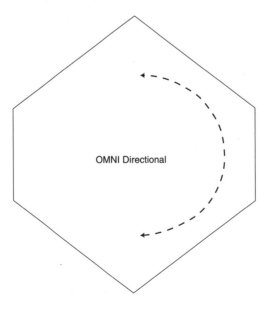

OMNI Directional

Due to the site's capability to receive signals from all directions, transmissions from neighboring sites may interfere with the site's reception. When interference reaches unacceptable levels, the site is usually sectorized to eliminate the capability to receive interfering information. Sectoring may also come into play when the site becomes so congested that the omnidirectional antennae cannot support the operations.

Sectorized Cell Coverage

Directional (sectorized) sites use reflectors positioned behind the antenna to focus the coverage area into a portion of a cell. Coverage areas can be customized to the needs of each site, as shown in Figure 4-12, but the typical areas of coverage are as follows:

- Two sectors using 180-degree angles
- Three sectors using 120-degree angles
- Six sectors using 60-degree angles

At least one transmit antenna is used in each sector (one per radio frame) and two receive antennas provide space diversity for each sector of

Figure 4-12
Sectorized coverage

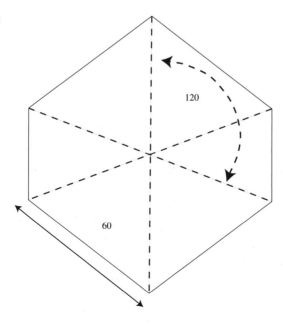

a two- or three-sectored site. One receive antenna is used in each of the 60-degree sectors, with neighboring sectors providing the sector diversity. This is an economics issue because of the number of antennas required.

Tiered Sites

This configuration places a low-power site in the same location with a high-power site. Mobiles change channels as they move across the boundary between the two in order to relieve congestion in the center, as shown in Figure 4-13. This configuration is used in all GSM and CDMA applications today. It is not supported on the older advanced mobile phone service (AMPS) and Digital AMPS (DAMPS) configurations, but may be used in newer implementations of AMPS and DAMPS. Each sector requires its own access/paging control channel to manage call setup functions. Voice traffic in each sector is supported by radios connected to antennae supporting that sector.

Reuse of Frequencies

Frequency reuse enables a particular radio channel to carry conversations in multiple locations, increasing the overall capacity of the communications systems. Within a cluster, each cell uses different frequencies; however, these frequencies can be reused in cells of another cluster.

Figure 4-13
Tiered-cell coverage

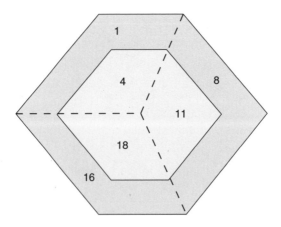

One centralized radio site with 300 channels can have 300 calls in progress at any one time. The 300 channels can be divided into four groups of 75 channels and still provide 300 calls at once. Dividing the service area into 16 sections called cells enables each cell to use one of the four groups of channels, increasing the call-carrying capacity of the system by a value of four (1,200 calls at one time).

The service area can be continually divided into smaller and smaller cells to obtain greater call-carrying capacity, increasing the number of calls by a factor of four with each division. The limit on how many cells can be used is determined by

- The cost infrastructure at each cell
- Processing power of the switch that controls the system
- Minimum power output at each site

Allocation of Frequencies

The allocation of frequencies based on the first cellular arrangement of *advanced mobile phone services* (AMPS) was designed around 666 duplex channels. The frequency ranges were allocated in the 825- to 845-MHz and 870- to 890-MHz frequency bands. In each band, the channels use a 30-kHz separation, and 21 channels are allocated to control channels. Figure 4-14 is a representation of the channel allocation.

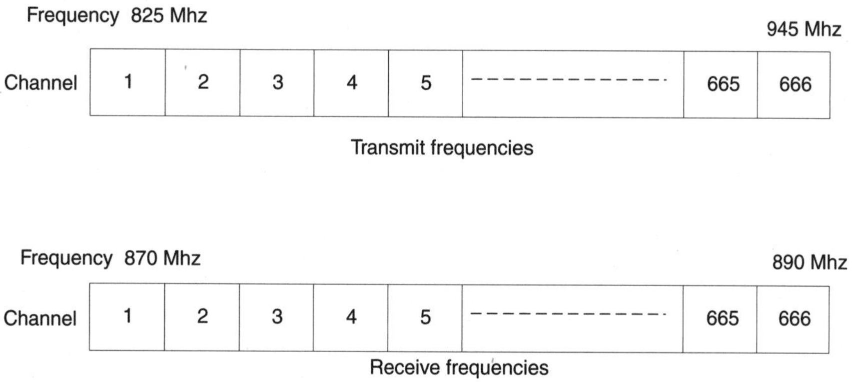

Figure 4-14
Frequency allocation for cellular

Frequency 825 Mhz   945 Mhz

| Channel | 1 | 2 | 3 | 4 | 5 | ------------ | 665 | 666 |

Transmit frequencies

Frequency 870 Mhz   890 Mhz

| Channel | 1 | 2 | 3 | 4 | 5 | ------------ | 665 | 666 |

Receive frequencies

In the U.S., the FCC approved licenses for two operators of the cellular service: the wireline carrier and the non-wireline carrier. The frequencies were split equally between the wireline and non-wireline operators. This meant that only half of the channels were available to each carrier and two sets of control channels were required.

Four signaling paths are used in the cellular network to provide signaling and control, as well as voice conversation. These can be broken into two basic function groups:

- Call setup and tear down
- Call management and conversation

Establishing a Call from a Landline to a Mobile

From a wired telephone, the local exchange office pulses out the cellular number called to the MTSO over a special trunk that connects the telephone company to the MTSO. The MTSO analyzes the number called and sends a data link message to all paging cell sites to locate the unit called. When the cellular unit recognizes the page, it sends a message to the nearest cell site. This cell site then sends a data link message back to the MTSO to alert the MTSO that the unit has been found. This message further notifies the MTSO which cell site will handle the call.

Next, the MTSO selects a cell site trunk connected to that cell and sets up a network path between the cell site and the originating CO trunk carrying the call. This is shown in Figure 4-15.

The MTSO is now also called the *Mobile Switching Center* (MSC). It is the controlling element for the entire system. The MSC is responsible for

- All switching of calls to and from the cells
- Blocking calls when congestion occurs
- Providing necessary backup to the network
- Monitoring the overall network elements
- Handling all the test and diagnostic capabilities for the system

This is the workhorse of the cellular system. The MSC relies on two different databases within the system to keep track of the mobile stations in its area.

Figure 4-15
Call establishment

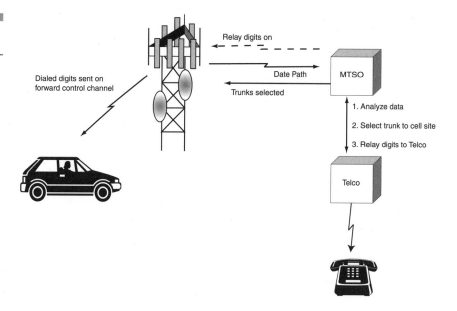

The first of the two databases is called the *Home Location Register* (HLR). The HLR is a database of all system devices registered on the system and owned by the operator of the MSC. These are the local devices connected to the network. The HLR keeps track of the individual device's location and stores all the necessary information about the subscriber. This includes the name, telephone number, features and functions, fiscal responsibilities, and the like.

The second database is called the *Visiting Location Register* (VLR), which is a temporary database built on roaming devices as they come into a particular MSC's area. The VLR keeps track of temporary devices while they are in an area, including the swapping of location information with the subscriber's HLR. When a subscriber logs onto a network and it is not home to that subscriber, the VLR builds a data entry on the subscriber and tracks activity and feature usage to be consistent for the user.

Another set of databases is used by the MSC in some networks. These are two separate and distinct functions called the *Equipment Inventory Register* (EIR) and the *Authentication Center* (AuC). They are very similar to the databases used in the GSM networks and keep track of manufacturer equipment types for consistency. The Authentication Center is used to authenticate the user to prevent fraudulent use of the network by a cloned device.

Intersystem Handoff

As we described with a handoff between cell site to cell site, the handoff can also take place between switching systems. This is a bit more complicated. When a vehicular phone is in use and serviced by a MSC (the MTSO), but the vehicle is getting very close to the outreaches of the MSC service area, something has to happen to prevent the caller from being cut off. The graphic in Figure 4-16 shows this occurring.

A message is sent out via the IS-41 network to the target MSCs in the area in which the caller is headed. The MSC sends a quality of signal measurement message to the targeted MSCs (which can be one to many). They all begin to listen in on the frequency channel in use by the caller. After a short time (less than one second), the MSCs all respond back to the serving MSC with the results of the measurement. One of the target MSCs will be receiving the signal the strongest, so the serving MSC selects that one.

After selecting the new MSC to serve the call, the serving MSC sends out a facilities directive to the new MSC. This tells the new MSC to activate a trunk between the two MSCs (assuming a trunk exists). The current MSC now becomes the anchor MSC and the new MSC becomes the serving MSC for the call.

Handoff Completion

Once the facilities directive has been enacted and the new target MSC responds, it notifies the serving MSC that a trunk has been connected and a new voice channel has been set up in parallel for the incoming roamer. The serving MSC now sends a message to the vehicle through the cell site over the air, commanding it to move to the new channel at the new serving MSC. The vehicle then retunes its dial to the new air channel and the call continues uninterrupted.

Figure 4-16
The intersystem handoff process begins.

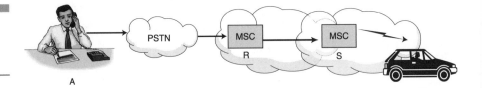

Hand Back

Occasionally, the caller will travel into a new MSC's serving area and get handed off. However, the way the geographical boundaries are set up, it is possible to be bounced back into the old serving area. As the vehicle starts to approach the boundary, the now serving MSC will issue a quality of service measurement initiative to the target MSCs in the area. Suprisingly, the anchor (the original MSC) is the one receiving the signal the strongest, as shown in Figure 4-17. Now the new serving MSC issues a facilities directive to the target MSC (the anchor) and tells it to set up a voice channel in parallel for the caller.

Once the MSC sends back its acknowledgment, the serving MSC will command the mobile to move to the new channel that has been reserved for it in the anchor MSC's serving area. Immediately upon the command and the mobile moving to the new voice channel, a facilities directive to release the trunk connection between the MSCs is issued and the trunk is taken down. This prevents what is called the shoelace problem.

Handoff to a Third MSC

Let's assume that the vehicle did not bounce back into the anchor's serving area, but continued eastward as shown in Figure 4-18. Now as the vehicle moves closer to the new area, the current serving MSC receives a message

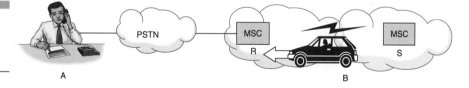

Figure 4-17
The handback prevents a shoelace problem.

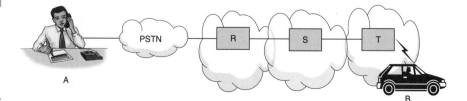

Figure 4-18
Handoff to a third party enables the caller to continue talking without being cut off.

back from the cells that the signal is getting weak. The now serving MSC issues a directive for quality of signal measurement to a new group of MSCs in the distance. This time a new MSC (labeled T) sends back an appropriate message to the serving MSC (labeled S) that the signal is getting strongest.

S will now issue a facilities directive to T to connect a trunk between S and T and to set up the parallel channel for the vehicle. Upon receipt of the acknowledgment, S will send a command to the vehicle to change channels to the new channel reserved by T. The mobile then retunes to that frequency and the call continues. Note that we now have three MSCs involved in the connection; R is the anchor, S is the intermediate node, and T is now the serving MSC.

The call can continue to be connected as long as trunks are available. If by chance S and T had no connection, things would be different. If a trunk was also between R and T, then S would have been removed from the call altogether and R, the anchor, would have a trunk-to-trunk connection to T directly. This is possible in some of the reciprocal networks used by the various players.

Seamless Networking with IS-41 and SS7

The whole intent of the IS-41 and SS7 interfaces is to enable the wireless carriers to communicate transparently and seamlessly between and among each other. Moreover, with SS7 interfaces to the wireline networks, calls can come into or exit the wireless networks flawlessly, as shown in Figure 4-19. This has been ongoing since 1994 (give or take) and seems to be moving quite well. In the event the wireless carriers do not have physical interfaces to the wireless providers or to the telephone companies, they can use the services of the *Independent Telecommunications Network* (ITN) to provide these services as a service bureau, for a fee. The interconnection between the networks provides industry standards-based internetworking. Moreover, the IS-41 and SS7 links are geared toward interfaces with *Local Number Portability* (LNP) switches. As LNP rolls out to the wireless suppliers, they will have to modify their networks to be AIN-1 compatible. Either they will provide the upgrades or use the service bureaus, for a fee. The point of this is when a wireless number is ported out to the wireline networks, then the wireless carriers will have to do the database dips and

Figure 4-19
Seamless networking
across various
provider networks

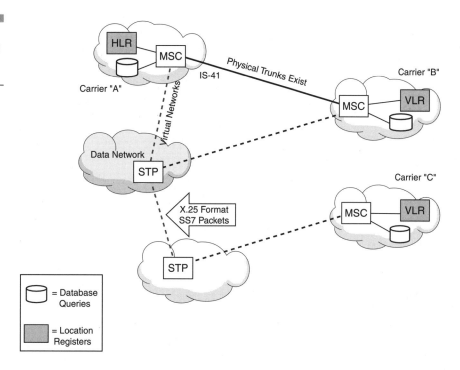

determine that an NXX has been ported. Then the NXX query will reveal
that the number has been ported. The wireless providers have still been
building their networks and are not complete in the renovations.

Automatic Roaming

The IS-41 and SS7 provide many of the features that will be required for
LNP, even though modifications will be necessary. First, automatic roaming
helps to define financial responsibility between the carriers for their users.
Secondly, it helps to provide the seamless interfaces between the wireless
systems and the SS7 interfaces between the wireline and wireless carriers.

As the system takes place, the automatic roaming enables the discovery
of a roamer to learn the identity of the current serving CO or visited system
to the home system.

A profile is established for the roamer in the visited system, allowing
the network to find the end user. This is how the wireless providers handle

location portability (of sorts). However, when asked, many of the wireless providers indicate that if a customer is going to be in an area for extended periods of time, they will take them off their current home system and put them into a new system. So much for transportability!

Lastly, the automatic roaming enables the setup for delivery for the calls. The wireless suppliers have been moving toward transparency through their bill-and-keep arrangement. In reciprocal billing, they assume that the calls cost an equal amount of money for the carriers to originate or terminate. Therefore, between wireless carriers, they do not charge each other for terminating minutes. The wireline companies, on the other hand, are charging for terminating minutes at approximately $.025 to $.03 per minute. Thus, the carriers are looking for a way to get away from the reciprocal billing and get to a bill-and-keep arrangement with the wireline providers.

Personal Communications

No one can ignore the changes that have taken place in the industry since the inception of the *personal communications services* (PCS). The entire industry has changed, making communications an integral component of our everyday life. What was always considered as frivolous (or for the affluent only) in the past has become commonplace today. One only has to drive down the road and see the other drivers in surrounding cars to recognize that personal communications is a reality.

Moreover, people in all walks of life have accepted the use of personal communications services. Look around to see where the personal communications systems have penetrated. What was once a business service is now an everyday component of students, businesspeople, and regular everyday people. The personal communicator, the wireless telephone, has become an indispensable tool for almost every occupation. The days of having a simple pager are quickly becoming obsolete as the personal communicator becomes more affordable and more competitively priced. Not too long ago, the industry was taken aback by the prospect of everyone having a personal telephone set. Today, this concept is closer to reality than ever before. As a matter of fact, the use of PCS has grown so much that now the vogue issue is to discourage people from using their phones. How often have you sat in a movie theater only to have a phone ring and the owner carry on a lengthy conversation? Restaurants have been inundated with so many complaints from other diners that they now are posting a notice prohibiting the use of wireless telephones in the dining area. This is a hot button for many other groups. The movement is to start banning the use of cell phones[1] in airports if they can get away with it. Moreover, local authorities throughout the U.S. are becoming extremely paranoid of these telephones because of the widespread use by drivers operating their vehicles and getting in accidents (causing them in many cases). The conversation is distracting and the driver is unable to use both hands on the wheel while holding a phone to his

---

[1] The term cell phone is used generically to include wireless sets and personal communicators.

or her ear. These occurrences are a common event. We even have people taking calls now on their personal communications device when they are in hospitals and churches (during services). I recently heard a phone ringing from a confessional. (Talk about long distance!)

School students are now the targets for the use of these devices, especially because the phones have dropped in price and appear cool in the social aspects among teens. Many students have gone to the designer phones so that they can be socially accepted.

This communicator is not a broadband device; however, it is the precursor for future devices. The newer ones will bring more bandwidth and more sophistication to our everyday lives. We have to merely walk before running! We will see high-speed communications moving to the handheld devices in the next few years, including data transfers at 170 Kbps in *general packet radio systems* (GPRS), at 2 Mbps in *third-generation* (3G) wireless systems, and then up to 64 Mbps for real-time video conferencing in the future. The timing of these changes is rapidly imploding. Already, many of the organizations supporting the PCS architecture have announced that they would introduce 3G wireless solutions in 2001–2002, yet commercial availability may take longer. The reason is that the push to 3G has encountered a minor slowdown. This slowdown is due in part from what everyone perceives as a lack of demand. Additional delays in regulation, spectrum availability, and technological interoperability have also compounded the targeted dates. Suffice it to say though, PCS is ready to roll as soon as these issues are resolved.

Personal communications systems have evolved from the wireless cellular and GSM networks, so not much excitement surrounds this topic. However, whereas the original wireless networks were built on an analog-networking standard, the PCS architectures are built on digital transmission systems. Therefore, several different approaches are used to deliver the capacities at the pricing models today. We will discuss the evolution of these wireless systems in their capability to propel us into the new millennium as the communicator of the future.

Current Cellular Standards

Different cellular systems employ different methods of providing multiple access to the system. The traditional analog cellular systems, such as those based on the *Advanced Mobile Phone Service* (AMPS) and *Total Access Communications System* (TACS) standards, use *Frequency Division Multiple Access (*FDMA).

FDMA

FDMA channels are defined by a range of radio frequencies, usually expressed in a number of kilohertz (kHz) within the radio spectrum. For example, the original AMPS systems use 30 kHz slices of spectrum for each channel. Narrowband AMPS (NAMPS) uses only 10 kHz per channel. TACS channels are 25-kHz wide. With FDMA, only one subscriber at a time is assigned to a channel. No other conversations can access this channel until the subscriber's call is finished or until the original call is handed off to a different channel by the system.

TDMA

A common multiple access method employed in new digital cellular systems is *Time Division Multiple Access* (TDMA). TDMA digital standards include

North American Digital Cellular (called IS-54 or D-AMPS), the European systems called *Global System for Mobile Communications* (GSM), and the Japanese systems called *Personal Digital Cellular* (PDC).

Digital Systems

Because FDMA uses the entire 30 kHz channels for one telephone call at a time, it is obviously wasteful. FDMA, an analog technique, can be improved a little by using the same frequency in a *Time Division Duplex* (TDD) mechanism. In this mechanism, one channel is used and time slots are created. The conversation flows from A to B and then from B to A. The use of this channel is slightly more efficient. However, when nothing is being sent, the channel remains idle. Because digital transmission introduces better multiplexing schemes, the carriers want to get more users on an already strained radio frequency spectrum. We saw how the FDMA slotting was laid out in the first chapter introducing the RF concepts.

Enhanced security and reduced fraud can be addressed with digital PCS. Again, this appears to be a win-win for the carrier for the following reasons:

- Less costs
- More users
- Better security
- Less fraud

The revenues, however, have been equally suspect in this industry. Over the past few years, the revenues from operations have been steadily dropping, partially due to competition and partially due to mass production efficiencies. Culminating this is the dropping cost per minute on long distance. This creates a conundrum for the carriers who are in need of finances for infrastructure expansion, acquisition of property for sites, and the bidding on spectrum. With steadily declining revenues per user, the need is to get more users. This becomes a catch-22 situation. Figure 5-1 is a representation of the sliding scales we see in the revenues. The figure represents the average revenue for voice-only and the microbrowser (handheld sets that can access the Web) sets on a global basis. The report shows a steady decline per year on the revenues per unit from the services. This also means that unless broadband and data services are added, the carriers will not be able to stem the decline. They must introduce new services and broadband applications in order to generate new life and revenue streams.

Figure 5-1
Average revenues
falling steadily over
the next few years
(Source: Ovum)

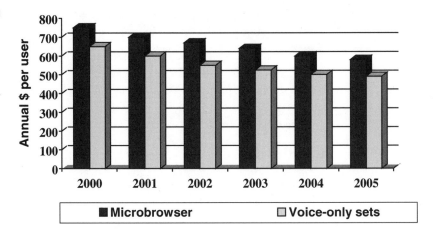

Once the decision was made to consider digital transmission, the major problem was how, what flavor to use, and how to seamlessly migrate the existing customer base to digital.

The digital techniques available to the carriers for PCS are

- TDMA
- CDMA

The two choices that we typically saw in this arena were the use of digital cellular or PCS. Digital cellular uses the 800- to 900-MHz frequency bands, whereas PCS operates in the 1,800- to 1,900-MHz frequency bands. Carriers and manufacturers are using a mix of these systems. The various means of implementing these systems have brought about several discussions regarding the benefits and losses of each choice. The real issue is that both exist and can be used differently.

Digital Cellular Evolution

Each move to digital requires newer equipment, which means capital investments for the PCS carriers. Moreover, as these carriers compete for radio frequency spectrum, they have to make significant investments during an auction from the FCC. This places an immense financial burden on the carriers before they even begin the construction of their networks.

Time Division Multiple Access (TDMA)

TDMA uses a time division-multiplexing scheme where time slices are allocated to multiple conversations. TDMA was introduced earlier in Chapter 3 "Access Techniques for Radio-Based Systems" during the discussion of access methods. Multiple users share a single radio frequency without interfering with each other because they are kept separate by using fixed time slots. The current standard for TDMA divides a single channel into six time slots. Then three different conversations use the time slots by allocating two slots per conversation. This provides a threefold increase in the number of users on the same radio frequency spectrum. Although TDMA deals typically with an analog to digital conversion using a typical pulse code modulation technique, it performs differently in a radio transmission. *Pulse Coded Modulation* (PCM) is translated into a Quadrature phase-shift keying technique thereby producing a four-phased shift, doubling the data rate for data transmission. Figure 5-2 shows the typical time slotting mechanism for TDMA.

The wireless industry began to deploy the use of TDMA back in the early 1990s when the scarce radio frequency spectrum problem became most noticeable. The intent was to improve the quality of radio transmission as well as the efficiency usage of the limited spectrum available. TDMA can correctly increase the number of users on the RF by three times. However, as more enhanced PCS techniques such as micro- and picocells are used, the numbers can grow to as much as a 40-fold increase in the number of users on the same RF spectrum. One can see why it is so popular.

TDMA has another advantage over the older analog (FDMA) techniques. Where the analog transmission across the 800-MHz frequency band (in North America it is 800 MHz, in much of the world it is 900 MHz) supports

Figure 5-2
TDMA for PCS

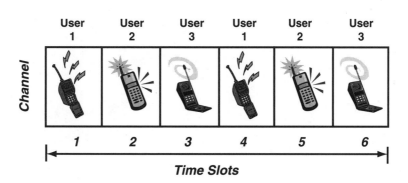

one primary service, that being voice, the TDMA architecture uses a PCM input to the RF spectrum. Therefore, TDMA can also support digital services for data in increments of 64 Kbps. The data rate can support from 64 Kbps to hundreds of megabits/second (120 Mbps). These rates, of course, will develop over time, but it is exciting to think that upwards of 120 Mbps could be possible in the future. Of course for now, many users would be happy with the 64-Kbps options.

This is an attractive opportunity for the carriers who are developing PCS and digital cellular using the industry standards. The two standards still in use today include IS-54, which is the first evolution to TDMA from FDMA for digital cellular, and IS-136, the latest and greatest technology for 1,900-MHz PCS service. When the IS-136 service was introduced, the addition of data and other services were introduced. These include the *short message service* (SMS), caller-ID display, data transmission, and other service levels. Using a TDMA approach, carriers feel that they can meet the needs for voice, data, and video integration for the future.

Penetration

From a global perspective, the use of TDMA for digital cellular and PCS has been very popular. This is evidenced internationally; GSM (see Chapter 6 "Global Services Mobile Communications"), the Japanese *Personal Digital Cellular* (PDC) networks, and the U.S.-based digital networks all primarily use TDMA as their infrastructure. Figure 5-3 is a graph of the penetration in 1999 for the various services as compared with CDMA on the basis of 465 million subscribers worldwide.

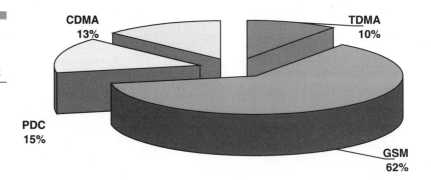

Figure 5-3
1999 approximate penetration of the various technologies

CDMA 13% TDMA 10% PDC 15% GSM 62%

The U.S. cellular market showed a penetration point of 31 percent in 1999. However, industry researchers expect that this number will grow to an overall penetration of 75 percent by 2007. This is shown in Figure 5-4.

Along this same line, we see that TDMA will continue to be popular although waning slightly over the future. It is estimated that in 2005 the numbers will shift from a total overall penetration of 87 percent in 1999; CDMA and third-generation wireless CDMA (3G-CDMA) will move up in penetration to approximately 40 percent globally on a population of 1.8 billion users. This is shown in Figure 5-5.

Figure 5-4
The U.S. cellular overall market penetration

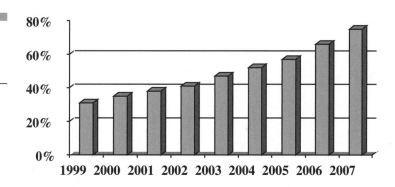

■ **Cellular Penetration by year**

Figure 5-5
Shifts in 2005 for TDMA and CDMA

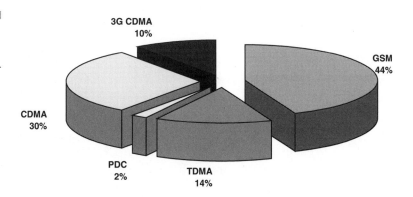

The CDMA Cellular Standard

Using *Code Division Multiple Access* (CDMA), unique digital codes, rather than separate radio frequencies or channels, are used to differentiate subscribers. The codes are shared by both the mobile station (cellular phone) and the base station, and are called *pseudo-random code sequences*. All users share the same range of radio spectrum.

In cellular telephony, CDMA is a digital multiple access technique specified by the *Telecommunications Industry Association* (TIA) as IS-95. Back in March 1992, the TIA created the TR-45.5 subcommittee. This committee's charter was to develop a spread-spectrum digital cellular standard. In July of 1993, the TIA gave its approval of the CDMA IS-95 standard.

IS-95 systems divide the radio spectrum into 1.25-MHz wide carriers. One of the unique aspects of CDMA is that although the number of phone calls that a carrier can handle is certainly limited, it is not fixed. Instead, a variable number of calls can be handled at the same time in the same frequency carrier. The capacity will depend on many other factors.

CDMA Development Group

The *CDMA Development Group* (CDG) is a nonprofit trade association formed in 1994 to foster the worldwide development, implementation, and use of CDMA. The primary activities include the technical development of features and services, public relations, education and seminars, regulatory affairs, and international development. Their involvement in the growth in popularity of CDMA systems has been exponential. Figure 5-6 is a graph of the comparison of growth for CDMA and TDMA over the past year.

CDMA-PCS

What is CDMA? CDMA is a spread spectrum technology, which means that it spreads the information contained in a particular signal over a much greater bandwidth (in this case the 1.25 MHz). IS-95 uses a multiple access spectrum spreading technique called *direct sequence* (DS) CDMA. *Direct sequence spread spectrum* is abbreviated as DSSS.

Figure 5-6

Penetration of digital cellular market by type (excludes GSM)

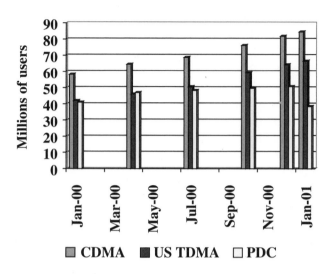

Millions of users

90
80
70
60
50
40
30
20
10
0

Jan-00 Mar-00 May-00 Jul-00 Sep-00 Nov-00 Jan-01

■ CDMA   ■ US TDMA   □ PDC

During a conversation (a cellular call), each user is assigned a binary, direct sequence code. The DS code is a signal generated by linear modulation with wideband *pseudo-random noise* (PN) sequences. Consequently, DS CDMA needs to use a much wider signal than other technologies. Wideband signals reduce interference and enable one-cell frequency reuse. No time division occurs; all users use the entire carrier all of the time.

CDMA cell coverage is contingent upon the way the system is designed. The three primary system characteristics include

■ Coverage

■ Quality

■ Capacity

These parameters must be balanced to achieve the highest level of system performance. In a CDMA system, these three characteristics are tightly interrelated. Higher capacity might be achieved through some degree of degradation in coverage and/or quality. Because these parameters are all intertwined, operators cannot have the best of all worlds: three times wider coverage, 40 times capacity, and CD-quality sound.

The 13-Kbps vocoder provides better sound quality, but reduces system capacity as compared to an 8-Kbps vocoder. Qualcomm engineered the use of both 8- and 13-Kbps vocoders. Motorola also uses many of these technologies in the operation of their systems. The best balance point may differ from one cell to another. Sites in dense downtown areas may trade off coverage for increased capacity. However, at the outer edges of a system where we expect fewer users, capacity might be sacrificed for the size of the coverage area.

Voice activity detection is another variable that helps to increase the capacity of a CDMA system. IS-95 CDMA takes advantage of voice activity gain through the use of variable rate vocoders. We must understand that in a typical phone conversation a person is actively talking only about 25 to 35 percent of the time. The difference is spent listening to the other party or is quiet time when neither party is speaking. The principle behind the variable rate vocoder is to have it run at high speed, providing the best speech quality, only when voice activity is detected.

When no speech is detected, the vocoder will drop its encoding rate because no reason exists to have high-speed encoding for silence. The encoded rate can drop from 13 or 8 to 4, 2, or even 1 Kbps. The variable rate vocoder frees up channel capacity in this case and only uses the higher rates as needed. Because the level of interference created by all of the users directly determines system capacity, and voice activity detection reduces the noise level in the system, capacity can be maximized.

More Sophisticated Vocoders

PCM is the vocoder standard used in wireline systems. It is simple, but not very efficient. It does produce high-quality sound using the 64-Kbps standard that wireless operators would like to match. Wired communications still uses PCM, despite its inefficiencies because bandwidth is cheap using fiber optic cable and/or microwave links.

Conversely, wireless vocoders are bandwidth constrained. Several types of vocoding standards currently exist, offering operators the choice between higher capacity and better voice quality. Initially, CDMA systems used the 8-Kbps variable rate speech vocoder, revision IS-96A. The vocoder transmits 8 Kbps of voice information at 9.6 Kbps when overhead and error correction bits are added. As a general rule, higher vocoder bit rates provide a more precise representation of a voice signal. However, older, less sophisticated vocoder designs may be unable to match the voice quality of these newer vocoders despite a higher bit rate.

The CDMA vocoder also increases call quality by suppressing background noise. Any constant noise (road noise and white noise) is eliminated.

The vocoder interprets constant background sounds as noise, which does not convey any intelligent information, and removes as much as possible. This greatly enhances voice clarity in noisy environments, such as the inside of cars, airports, planes, or restaurants (if they still let you use the phone in these places).

Capacity Improvements

Capacity is a function of the number of channels installed in a specific coverage area. Using an example of what we have already seen in an AMPS environment, let's look at the possibilities of the use of CDMA in the same geographic area. In an earlier chapter, we looked at the possibility of using the tiered and the sectorized cells using AMPS. This showed where we could use upwards of 300 channels in the same physical area of a cell. However, this also required extensive equipment upgrades. One CDMA carrier requires 1.25 MHz of bandwidth. Because three-sector AMPS use a seven-cell reuse pattern, this example will spread the 1.25 MHz across seven cell sites. Each cell site would then lose 180 kHz of spectrum (1.25 MHz ÷ 7 = 0.180 MHz). Thus, a total of six AMPS channels must be removed from each cell site (180 kHz ÷ 30 kHz/AMPS channel = 6). The result means that 42 AMPS channels must be removed in order to support one CDMA carrier.

Unlike AMPS, CDMA can use the same 1.25 MHz in all three sectors in each of the seven cells. Many of the system designs from Motorola, for example, support 18 effective traffic channels per sector in a three-sector system. This provides 54 effective channels per cell. Given the seven cells, CDMA supports 378 channels. Hence, in this example, CDMA achieves capacity gains of nine times that of AMPS (378 ÷ 42 = 9). We continually hear of between eight to tenfold increases with CDMA. This example clearly points this out.

CDMA Benefits

When implemented in a cellular telephone system, CDMA technology provides many additional benefits to cellular operators and their subscribers. Table 5-1 summarizes the benefits of CDMA.

Table 5-1

Summary of
CDMA Benefits

Description	Benefit goes to
1. Capacity increases of 8 to 10 times that of an AMPS analog system	Cellular provider: The increased revenues and the same radio frequencies, helping service providers increase their profitability. CDMA spread spectrum technology can provide up to 10 times the capacity of analog equipment and more than 3 times the capacity of other digital platforms such as TDMA. With dual-mode phones, CDMA is also compatible with other technologies for seamless widespread roaming coverage.
2. Improved call quality, with better and more consistent sound as compared to AMPS system	Cellular provider and user: Improved overall performance and quality of voice communications. CDMA filters out background noise, cross talk, and interference delivering improved voice quality, greater privacy, and enhanced call quality. The CDMA variable rate vocoder engineered by Qualcomm translates voice into 0's and 1's. These vocoders operate at the highest translation rates possible (8 Kbps or 13 Kbps). This creates crystal clear voice and also maximizes your system capacity. Many of the compression schemes in use today operate at 5.3–6.3 Kbps, which have a tendency of distorting the voice conversation, and have trouble passing data communications. CDMA combines multiple signals and improves signal strength. This nearly eliminates interference and fading. Electrical noise (computer noise) and acoustic noise (conversations) are filtered out. This is possible by using narrow bandwidth corresponding to the frequency of the human voice.
3. Simplified system planning through the use of the same frequency in every sector of every cell	Cellular provider: Ease of design and installation. Less frequency coordination required and less interference inside cells and/or between cells. CDMA systems can be deployed and expanded faster and more cost-effectively. Because they require fewer cell sites, CDMA networks can be deployed faster.
4. Enhanced privacy	Cellular provider and user: Enhanced security means less worry. Less reluctance to use the system. CDMA uses a digitally encoded, spread spectrum transmission that resists eavesdropping. Designed with about 4.4 trillion codes, CDMA virtually eliminates cloning and other types of fraud.
5. Improved coverage characteristics, creating the possibility of fewer cell sites	Cellular provider: More coverage and less cost all at the same time, fewer calls dropped during handoff because fewer handoff conditions will be present. CDMA-patented soft handoff method of passing calls between cells sharply reduces the risk of disruption or dropped calls during a handoff. The process of soft handoff leads to fewer dropped calls as two or three cells are monitoring a call at any given time. CDMA spread spectrum

(continued)

Table 5-1

Summary of
CDMA Benefits
(continued)

signal also provides the greatest coverage, allowing networks to be built with fewer cell sites than other wireless technologies. Fewer cell sites reduce operating expenses, which results in savings to both operators and consumers.

6. Increased talk time for portables	User has less hassles and battery replacement/charging. Fewer calls will drop due to lack of power. Users can leave their phone on with CDMA. CDMA uses power control to monitor the power your system and handset need at any time. CDMA handsets typically transmit at the lowest power levels in the industry, enabling longer battery life, which results in longer talk time and standby time.
	CDMA handsets can also incorporate smaller batteries, resulting in smaller, lighter-weight phones. They are therefore easier to carry and easier to use.
7. Bandwidth on demand	User: Growing access and more useful bandwidth utilization in CDMA technology enable users to access a wide range of new services, including caller identification, short messaging services, and Internet connections. Simultaneous voice and data calls are also possible using CDMA technology. A wideband CDMA channel provides a common resource that all mobiles in a system utilize based on their own specific needs, whether they are transmitting voice, data, facsimile, or other applications.
8. Packetized data	CDMA networks are built with standard IP packet data protocols. Other networks require costly upgrades to add new packet data equipment in the network and will require new packet data phones. Standard CDMA phones already have TCP/IP and PPP protocols built into them.

CDMA Today

CDMA is the fastest-growing wireless communications technology today throughout the world. Currently, CDMA offers the fastest data rates for wireless data applications with up to 64 Kbps. This is a given, whereas the use of GPRS currently only offers 28.8 Kbps and promises 170 Kbps for the future. CDMA is also the technology of choice for many of the 3G products and services. A single CDMA standard with three modes provides flexibility for all operators to meet the growing demand for advanced voice and data services. Probably the most important concept to any cellular telephone system is that of multiple access, vis-a-vis simultaneous users can be supported. In other words, a large number of users share a common pool of radio channels and any user can gain access to any channel. (No user is permanently assigned to the same channel; instead he or she competes for the right to use the first available channel.) A channel can be thought of as a slice of the limited radio frequency spectrum temporarily allocated to someone's phone call. Multiple access defines how the radio spectrum is divided into channels and how channels are allocated to the users of the system.

Rationale Behind CDMA's Popularity

As the use of radio frequency spectrum continued to put pressure on this limited resource, the manufacturers of systems and regulators were searching for some way to share spectrum among multiple users. Furthermore, the sharing is compounded by the need to secure information while on airwaves. These pressures have led to the use of spread spectrum radio. The spreading portion of these systems using a chip set coded for a specific transmitter to receiver system uses multiple frequency (called hopping) as one method; or its technique of creating a coded chip set is used. Either way, both of these services are designed to spread as much energy over a broader range of frequencies to enable less airtime on a specific bandwidth and to ensure the integrity of the information being sent. The technique for the spread spectrum service is called *Code Division Multiple Access* (CDMA) with *direct sequence spread spectrum* (DSSS). Many of the PCS carriers have chosen CDMA DSSS as their coding of choice. The coded chip method was used in military applications and high noise areas where the risk of interference is high.

Because spread spectrum has been introduced at the commercial level, the FCC allocated spectrum in the 1.9-GHz range. The benefits of frequency hopping may be overshadowed by distance, utilization, and power constraints.

Soft versus Hard Handoff

Traditional cellular systems use a hard handoff, whereby the mobile drops a channel before picking up the next channel. A soft handoff occurs when two or more cell sites monitor a mobile user and the transcoder circuitry compares the quality of the frames from the two receive cell sites frame by frame. The system can take advantage of the moment-by-moment changes in signal strength at each of the two cells to pick out the best signal.

To make sure that the best possible frame is used in the decoding process, the transcoder can toggle back and forth between the cell sites involved in a soft handoff on a frame-by-frame basis (if that is what is required to select the best quality of conversation).

These soft handoffs also contribute to high call quality by providing a *make before break* connection. How many times have we used the older systems and gotten cut off in the handoff process? This dropped call is a result of when the RF connection breaks from one cell to establish the call at the destination cell during a handoff. We hear this as a short disruption of speech with non-CDMA technologies. Narrow-band technologies compete for the signal. When cell B wins out over cell A, the user is dropped by cell A (hard handoff). With CDMA, the cells work together as a team to successfully achieve the best possible information stream even if it is shared among the cells. Eventually, cell A will no longer receive a strong enough signal from the mobile, and the transcoder will only be obtaining frames from cell B. The handoff will have been completed, undetected by the user. CDMA handoffs do not create the hole in speech that is heard in other technologies.

Some cellular systems also suffer from the ping-pong effect of a call being repetitively switched back and forth between two cells when the subscriber unit is moving along a cell border. At worst, such a situation increases the chance of a call getting dropped during one of the handoffs and at a minimum, causes noisier handoffs. CDMA soft handoff avoids this problem entirely. Finally, because a CDMA call can be in a soft handoff condition among three cells at the same time, the chances of a dropped call are greatly reduced. CDMA also provides for softer handoffs. A softer handoff occurs when a subscriber is simultaneously communicating with more than one sector of the same cell.

Over-the-Air Activation

Over-the-Air Activation is a feature that is key to the future business plans of many wireless operators. This feature, developed by the CDG, enables a potential cellular service subscriber to activate new cellular service without the intervention of a third party, such as an authorized dealer. A cellular user can activate a user-controlled feature and service.

One of the primary objectives of over-the-air activation is to provide a secure authentication key to a mobile station to facilitate the authentication process. *Authentication* is the process by which information is exchanged between a mobile station and the network for the purpose of confirming and validating the identity of the mobile device. A successful outcome to the authentication process occurs only when it is demonstrated that the mobile station and the network possess identical sets of shared secret data (keys).

The over-the-air activation feature consists of over-the-air programming of the *Numeric Assignment Modules* (NAMs), which are used to authorize cellular telecommunications service with a specific service provider. The feature incorporates an authentication key exchange agreement algorithm. This algorithm allows the network to exchange authentication key parameters with a mobile station. These parameters are used to generate the authentication key that is used to generate the shared secret data. The authentication key exchange agreement algorithm enhances security for the subscriber and reduces the potential for fraudulent use of cellular telecommunications service. This is similar to an encryption key used between computer systems. If you have the correct code (key), you can access information. If you do not have the appropriate key, you cannot read the data.

This feature alone has done more to substantially reduce distribution and activation costs for the providers and reduce frustration for the users. Other features can be activated by these specifications. For example, subscribers could gain easy access to other CDMA systems through automatic updates of roaming information; or a new feature can be turned on, without the user ever having to go to the cellular provider's location.

What About Data?

CDMA circuit-switched services provide asynchronous data and facsimile transmission using an *Interworking Unit* (IWU). The IWU provides the functions needed for the mobile equipment to communicate with fixed end equipment in a public network. This architecture adapts the air interface and landlines by providing retransmission protocols unique to the CDMA air interface, called the *Radio Link Protocol* (RLP), and rate adaptation to the landline modems.

Various data services are initiated as service options during call setup or anytime during a call. The service option negotiation process specifies whether the service option will be used for the primary or secondary traffic. Therefore, the user can switch between voice, data, and fax service simply by initiating and terminating the appropriate service options.

This has led to innovations by Motorola and others in the CDG to develop and support the *L* interface, TIA standard IS-687 PN3473. The *L* interface allows both Interworking Units and Data Gateways to communicate with any infrastructure equipment that also uses this interface.

Circuit Mode Asynchronous Data/Fax Rates

CDMA systems support synchronous and asynchronous data services that emulate a traditional dial-up modem connection to the PSTN. In the CDMA system, the modem is not in the subscriber unit, but in the network. This enables direct digital communications over the radio channel. It preserves the advantages of true digital transmission, eliminating the need to convert from digital to analog and then back to digital. Using this approach enables a subscriber unit to transparently communicate with any landline modem. The landline user modem can support any existing V-Series modulation techniques. Typically, a data rate of 9.6 Kbps is supported with the 8-Kbps vocoder and 14.4 Kbps with the 13-Kbps vocoder. Higher speeds have been endorsed and designed to support 64 Kbps.

Any of the transmission rates in the CDMA system can be used for data. To support efficient transmission, flow control is an integral part of the CDMA system. Thus, the air interface transmission rate need not match the landline rate. For example, the air interface rate can be higher than the landline rate to reduce delay. Similarly, the air interface rate can be lower than the landline rate to support more users having simultaneous traffic demands. The CDMA system also supports Group III fax standards, which operate from 9.6 to 14.4 Kbps.

Simultaneous Voice and Data

CDMA systems support the simultaneous transmission of voice and data. The two digital streams will be multiplexed on a frame-by-frame basis with voice being given priority over data to maintain voice quality. As higher data rate channels are introduced later, data throughput will increase. Several modes of operation including turning on and off voice service during a data call or adding data to a voice call in progress will be supported.

Packet Data Services

CDMA is defined as a lower level protocol. The CDMA system will support higher level protocols commonly used for data communications, such as TCP/IP. As the layers 1 and 2 of the OSI model, CDMA will carry the traditional IP packets. All CDMA sets today support the inherent passing of IP datagrams onto the channel. Simultaneous voice and data will be available in the case of packet data applications.

PCS Providers

These carriers seem to be the ones seeking a niche in the market. Because they offer services to their customers, they view themselves as an alternative to the ILEC wireline services and complementary at the same time. Depending on how the customer reacts, the PCS providers especially will move toward one position. In checking with several PCS suppliers, the argument to make the PCS telephone number the sole number came up. The providers offer the same features and functions as the wireline service providers such as

- Call forwarding (busy or don't answer)
- Voice messaging
- Three-way calling
- Caller ID
- Call transfer

The PCS providers are now saying that the customer can use the same number for the home number, the business number, and the traveling number. Why pay for two different lines and service offerings when you can do it all on one telephone (and number)? Their speech is somewhat convincing, but they fail to mention the cost of the airtime for receiving calls and the added cost for making local calls. However, they are now packaging these services in such a way that they are invisible.

The overall use of PCS will continue to put pressure on the wireline providers. Voice, data, and video will all be available from a broadband PCS device someday in the future. These providers will continue to invest in architecture to meet the future broadband demands and selectively serve niche markets. PCS will be one of the primary sources of service for the future.

Global System for Mobile (GSM)

In the early 1980s, analog cellular telephony systems were rapidly growing in the European marketplace, particularly in the United Kingdom, France, Germany, and Scandinavia. In each country, the providers had developed their own internal operating systems to support this new mobile communications revolution. Unfortunately, these locally developed country systems were incompatible from system to system and country by country. Obviously, this was the least desirable of all situations for the introduction of a new system, leaving operators and users equally dissatisfied. Something had to be done to create a unified approach to the wireless networking and communications systems to bring them into harmony.

In 1982, the *Conference of European Post and Telegraph* (CEPT) created a study group to analyze what could be done. This study group was named *Groupe Special Mobile* (GSM). Their charter was to develop a system that would work across the European market. The systems they proposed had to be capable of meeting certain criteria such as the following:

- Good quality speech
- Low cost for the equipment
- Efficient use of RF spectrum
- Capable of supporting the newer handheld telephones
- Transparent roaming capabilities
- ISDN compatible

The committees did their job well, and quickly endorsed the standards and specifications to create a special mobile communications system capable of working across the international boundaries that had heretofore been blocked. These specifications were handed over to the *European Telecommunications Standards Institute* (ETSI) in late 1989. In 1990, GSM specifications were published. This turnover happened relatively quickly, considering the fact that political and economic in fighting typifies standards setting in this industry. However, the setting of specifications led to the rollout of many GSM systems; in 1993, 36 networks were operational in 22 countries, signifying the rapid acceptance and development of a single, standards-based network. Now, over 300 operators in 133 countries have already endorsed and accepted the GSM specification for their local and national wireless networking standard.

The graph in Figure 6-1 shows the estimates that highlight the growth of the GSM industry around the globe. In North America, several carriers have introduced GSM in their networks, showing a unified approach to the worldwide standardization. In some cases, the GSM installations in North America use a derivative of GSM called PCS 1900.

Figure 6-1
Percentage of
subscribers on GSM
worldwide

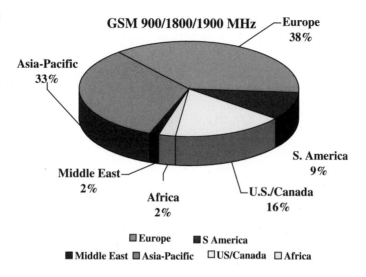

The graph shows the relative percentages of GSM users in the world by region. Table 6-1 is the summary of the actual subscribers of the various GSM technologies in numerical value. GSM amounts to approximately 58 percent of the world's wireless implementations spanning the 900-, 1,800- and 1,900-MHz bands. In North America, the United States, and Canada, approximately 109 million users span 17 operators in over 1,550 cities across the continent. The combined global *Time Division Multiple Access* (TDMA) and GSM subscriber base on five continents was approximately 465 million by the end of the year 2000. This combined subscriber base represents over 70 percent of the estimated 655 million global wireless subscribers. By joining forces in the development of interoperability standards for global roaming, carriers will be able to offer seamless international service to their customers.

Table 6-1

Market Share by
Region of GSM
900/1800/1900
MHz

Region	Number of Users in Millions
Europe	248.4
South America	62.4
U.S. and Canada	109.7
Africa	15.9
Middle East	10.5
Asia-Pacific	219.5

Change Is Underway

The telecommunications world is changing as the trends of media convergence, industry consolidation, Internet and *Internet Protocol* (IP) technologies, and mobile communications collide into one. This rapid evolution in technology will bring about significant change. With *third-generation* mobile Internet GSM technology, a radical departure is occurring from that which came in the first and even the second generations of mobile technology. Some of the changes include

- People will look at their mobile phone as much as they hold it to their ear. As such, 3G will be less safe than previous generations because television and other multimedia services tend to attract attention to themselves; instead of hands-free kits, we will need eyes-free kits!
- Data (non-voice) uses of 3G will be as important as, and very different from, the traditional voice business.
- Mobile communications will be similar in its capability to fixed communications, such that many people will *only* have a mobile phone.
- The mobile phone will be used as an integral part of the majority of people's lives. It will not be an added accessory, but a core part of how they conduct their daily lives. The mobile phone will become akin to a remote control or magic wand that allows people do what they want, when and where they want.

3G terminals will be significantly more complex than today's GSM phones because of the need to support video, more storage, multiple modes, new software and interfaces, better battery life, and so on. The largest inhibitor to new services such as *Wireless Application Protocol* (WAP) and *High Speed Circuit Switched Data* (HSCSD) has already proven to be a lack of handsets. Every stage in the data evolution path for GSM from today to 3G requires a new handset. Once again we see that terminals are mission critical and their timely volume availability will be a critical factor in determining when 3G is a success.

GSM Concept and Services

Planning a GSM specification was not a simple task for the developers. They initially wanted to overcome all the pitfalls of the older telecommunications technologies and networks. One area that the developers sought to address was the use of a 64-Kbps channel capacity for ISDN integration. However, when using a wireless communications system, the in-band signaling and the robbing of bits were not conducive to delivering a full 64-Kbps channel. The basic architecture of GSM is to support one primary service, that being telephony from a mobile perspective. No matter what architectural model is used, speech-encoding techniques are used to transmit the voice as a string of digital 1's and 0's. GSM accommodates an emergency response system similar to the North American 911 services, whereby the local emergency response agency is notified via a three-digit number. Other services are automatically included in the operation of a GSM service and network, such as

- Circuit-switched data
- Packet-switched data
- Voice
- *Short message services* (SMS)
- ISDN
- Facsimile using an ITU Group 3-fax service

The GSM Architecture

The network is comprised of several components, not unlike the analog cellular networks of old. The functionality of each of the components describes the overall complexity and the degree of robustness built into the network. The pieces are shown in Figure 6-2, which depicts the layout of the service elements and their interrelationship with each other. The figure shows the component, whereas the actual function that each piece performs is described in the following section. Three major components come into play with the GSM network: the mobile unit, mobile base station, and mobile switching system.

- **The mobile unit** The mobile unit, or mobile station, consists of the mobile telephone unit and typically a smart card called the subscriber interface module or *Subscriber Identity Module* (SIM). The SIM provides mobility for the individual so that a user can roam seamlessly and have all the services contracted for, regardless of the end user terminal device. By inserting the SIM into the set, the set takes on the personality of the end user. The user is then able to make and receive

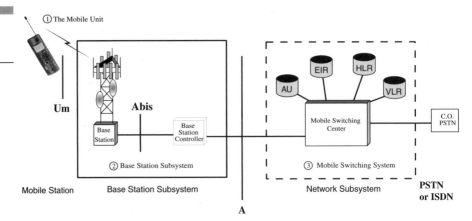

Figure 6-2
Architecture of the
GSM network

calls and receive the features allowed by contract, even if the set is a temporary one.

The mobile unit has a specific identifier called the *International Mobile Equipment Identity* (IMEI). The SIM contains the *International Mobile Subscriber Identity* (IMSI) used to identify the individual subscriber to the mobile system. Furthermore, the IMSI contains authentication information in the form of an encrypted key. Any other pertinent information required by the systems operator is also contained in the IMSI. Because the two sets of identifiers are unique yet independent, mobility is assured for the individual.

■ **The base station subsystem**   The base station subsystem consists of two additional parts. The first part is the *Base Transceiver Station* (BTS). The second part is the *Base Station Controller* (BSC). These devices communicate between the components of the system or to disparate manufacturer's equipment through a standards-based interface (the Abis interface). The BTS is where the radio systems are located for the air interfaces to the subscriber mobile unit. Radio link protocols for GSM are used between the BTS and the mobile unit. The typical GSM air interface uses *Time Division Multiple Access* (TDMA) radio protocols.

The BSC manages the radio resources for at least one, but possibly many, base transceiver units. The BSC is also the interface between the mobile unit and the *Mobile Switching Center* (MSC). The combination of the BTS and the BSC are defined as the base station subsystem.

■ **The network subsystem**   At the core of the network subsystem is the MSC. It acts like the Class 5 Central Office of the *Public Switched*

Telephone Network (PSTN). Moreover, the MSC provides all the necessary switching and call processing functions for the mobile unit, as well as additional functions like authentication, mobile handoff, registration onto the network, and a group of database functions. The MSC also provides the interface to the backbone signaling networks for call setup and teardown through the CCS7 (SS7) networks.

Residing at the MSC is the *Home Location Register* (HLR), which is a database of the registered users to a specific network system. The HLR is the owner of the SIM for each subscriber of a specific network operator. When a user moves from one system to another, the *Visitor Location Register* (VLR) is enacted. The VLR is a temporary database of visiting devices in a system's area of operation. The VLR will notify the individual owner's HLR that the device is temporarily in a new location and that all requests for service can be handled by the VLR. Logically, only one HLR exists per GSM network, although it can be distributed across many locations. The VLR also contains selected information from the HLR to provide all the necessary call control information for the subscriber.

It should therefore be noted that the MSC does not play into the information on specific devices because the information resides in the databases (registers) rather than in the switching system. Additional registers are used for enhanced control, such as the *Authentication Center* (AuC) register and the *Equipment Inventory Register* (EIR). The AuC is an authentication server used to verify the user's specific information, such as password and authentication keys. The special keys and passwords are used in an over-the-air interface, so the authentication server validates the information and protects the information. The EIR is a database of the vendor-specific information for all the radio sets used on the network, by manufacturer, and by IMEI. If the set is stolen or suspected of fraud, its IMEI is flagged as invalid in the database, denying a user with a clone or stolen set from using the network.

The Air and Link Interfaces

Internationally, the ITU allocated radio frequency spectrum in the 890- to 915-MHz band for the uplink and in the 935- to 960-MHz band for the downlink in Europe. Figure 6-3 shows the frequency used in the specific up-and-down link structures. The uplink is from the mobile unit to the base station, whereas the downlink is from the base station to the mobile unit as shown in Figure 6-3. Because these same frequencies were already in use by the initial analog networks, the CEPT reserved the top 10 MHz of each band for the GSM network, which was still being developed at the time. This means that the GSM networks operate with an uplink capacity of 905 to 915 MHz and 950 to 960 MHz for the downlink. Over time the analog networks will decline, and the full 25 MHz in each band will be allotted to the GSM networks. This has yet to happen.

The Access Techniques Used

GSM, as stated earlier, uses TDMA in the air interface. In reality, GSM uses a combination of TDMA and *Frequency Division Multiple Access* (FDMA). Where a 25-MHz band is allotted for GSM, FDMA is used to break the 25-MHz spectrum down into a total of 124 carrier frequencies spaced at 200 kHz apart. One or more of these carrier frequencies is assigned to each of the base stations. From there, each of the carrier frequencies is subdivided in time division using TDMA. The basic element of the TDMA is a burst of data during a burst period of 0.577 ms. Eight burst periods are grouped into a frame. The TDMA frame is approximately 4.615 ms. This framing forms

Figure 6-3
The up-and-down link spectrum allocated

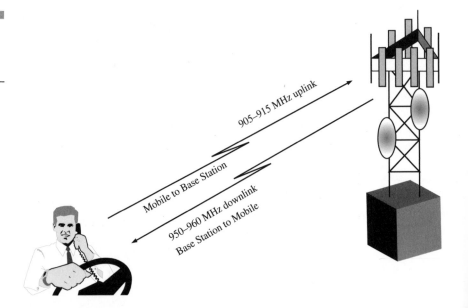

the basis of a logical channel. One physical channel is one burst period in a TDMA frame. The number and position of its corresponding burst period defines the channels. All definitions operate on a cycle of patterns that repeat every 3 hours.

Traffic Channel Capacities

A *traffic channel* (TCH) carries speech or data traffic. Traffic channels are defined by using groups of 26 TDMA frames called multiframes. A multiframe is 120 ms long (120 ms/26 frames/eight burst periods per frame). The multiframe is broken down into the following pieces:

- Twenty-four frames carry traffic.

- One frame carries a *Slow Associated Control Channel* (SACCH).

- One frame is unused.

Figure 6-4 shows the frame breakdown. The traffic channels are separated between the up-and-down link by three burst periods, so the mobile does not have to send and receive simultaneously. This simplifies the electronics used in the system.

GSM also has provisions for half-rate channels, although their implementation is not fully common. Half-rate TCHs will ultimately double the capacity of a system. To do this, half-rate speech coders will be used (using 7 Kbps of speech instead of the normal 13 Kbps).

Figure 6-4
TDMA frames and
multiframe allocation

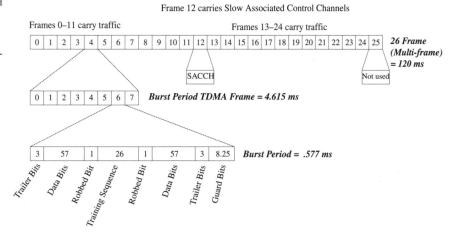

Control Functions

Many of the aspects of the GSM use common control channels for either idle mode or dedicated mobile channels. These common channels are used by idle sets to exchange signaling information between the base and the mobile in order to change from an idle to a dedicated mode. A dedicated channel is allocated to a specific mobile while engaged in conversation, whereas an idle mode is used when the mobile is not engaged in talk. Mobiles that are in a dedicated mode monitor all the surrounding base stations for handoff and other control information. To separate the common and dedicated channels, the common channels are defined in a 51-frame multiframe. This way a dedicated mobile can use the 26-frame multiframe for speech and traffic, yet can still listen to the monitor channels. Several different common channels are defined:

- **Broadcast control channel**   A constant broadcast of information regarding frequencies, frequency hopping patterns, and other downlink information.

- **Frequency control and synchronization channel**   Time synchronization pattern used to align the time slots for the cells. The cells use one of each of these channels allotted to channel slot number 0 in a TDMA frame.

- **Random access channel**   A slotted aloha channel used by the mobile unit to request access to the network.

- **Paging channel**   The paging channel is used to signal a mobile unit of an incoming call request.

- **Access grant channel**   This channel is used to allocate a stand-alone dedicated control channel to a mobile for signaling, usually following a request to access the network.

The Data Burst

Figure 6-4 shows the format of a TDMA frame. This figure also shows the organization of the data bursts. The normal burst carries data and signaling information. This burst lasts for 156.25 bit times, consisting of two separate 57-bit data patterns, a 26-bit training sequence for equalization, a robbed bit for forward synchronization in each block of data, three trailer bits at each end, and 8.25 bit times guard band. The 156.25 bits are transmitted in the allotted 0.577 ms, yielding a data rate of 270.833 Kbps.

Speech Coding Formats

When speech is to be coded onto GSM, it is a digital transmission system. Therefore, the analog voice is converted to digital before transmission. The normal telephony architectures use standard PCM techniques to digitally encode a voice signal at a data rate of 64 Kbps. However, 64 Kbps is too difficult to accomplish across the radio signals. A speech compression and coding technique using a form of *linear predictive coding* (LPC) produces a 13-Kbps speech pattern. The speech is actually divided into 20-ms samples, encoded as 260 bits each, yielding the 13-Kbps speech.

The Network Structured Protocols and Interfaces

When using GSM, a layered protocol approach occurs for the several interfaces and the protocols necessary to provide transparency across the medium. Figure 6-5 portrays the protocol stack at the different components of the network architecture. The protocols stack for the mobile station is in the leftmost portion of the figure. The TDMA protocol resides at the physical layer interface (the air interface). *Link Access Protocol Data for Mobile* (LAP-D$_m$) protocol, which is a derivative of the ISDN LAP-D data link layer, works at layer 2. The GSM layer (layer 3) is subdivided into three separate sublayers: the radio resources management sublayer, the mobility management sublayer, and the communications management sublayer.

As the information is passed between the mobile unit and the base transceiver station, the layers are similar, but only used from the TDMA layer up

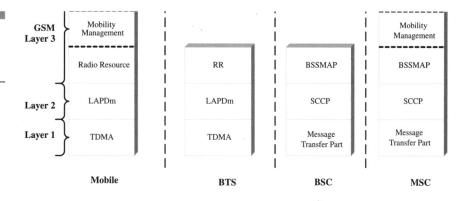

Figure 6-5
The signaling protocol stack for GSM interfaces

to the radio resource management sublayer. The upper sublayers of layer 3 are not used at the user to mobile interface. This is shown in the next part of the drawing in the second stack from the left.

The Abis interface is used between the base transceiver systems and the base station controller. At the protocol stack, the layer 1 protocol is the *Mobile / Message Transfer Part* (MTP), which acts at the bottom three layers of the typical OSI interfaces. The *Signal Connection Control Part* (SCCP) and the *Base Station System Mobile Application Part* (BSSMAP) combine to form the equivalent of an SS7 protocol stack. This is shown in the third stack from the left.

The BSSMAP is the equivalent of a TCAP, as shown in Figure 6-6, which is merely a comparison of the *Signaling System 7* (SS7) protocol stack.

The stack remains on the right-hand side of Figure 6-6, showing the Mobile Switching Center protocols as the signal is passed to the MSC. In this case, the mobility management and the communication management parts are shown on the upper layers of the stack.

These additional pieces include such added features and functionality in the protocol stack to support the following:

- **Radio resource management** This deals with link establishment and maintenance, the handoff between different base stations, and the

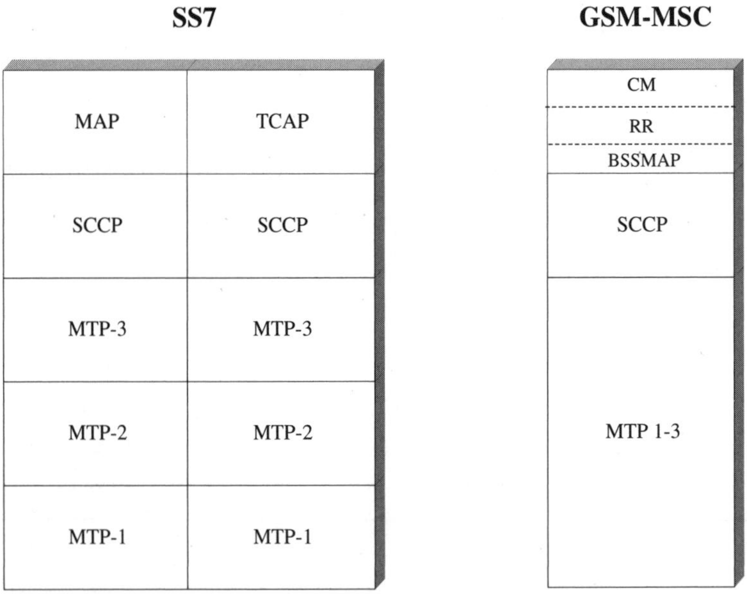

Figure 6-6
Comparing the TCAP and the GSM BSSMAP

coordination of the necessary spectrum used in a particular base station operation. This is also used in setting up the necessary channel assignment when a call is coming in or a page is initiated across the wireless interface.

■ **Mobility management**   This deals with the constant location updates between the HLR and VLR function. Also included in the mobility management are the authentication, equipment inventory, and security capabilities.

■ **Communications management**   This deals with the control of call setup, routing, and teardown functions. The call control function attempts to use a point of Q.931 for routing and delivery of calls. GSM protocols are similar but unique from the standard wired telephony network. This is the equivalent of the ISUP from an SS7 standpoint.

Some Thoughts on GSM

GSM brings a standardized platform for communications within the mobile environment. In North America, several different approaches were used to implement mobile and personal communications services. The rest of the world has lock-stepped into a global means of providing the transparency and seamless roaming capability needed to provide true communications spirit. While in Europe, a user can rent a telephone set (or buy one) and install the smart card into it. Immediately, the set takes on the identity of the individual installing the smart card. From billing information, features, and subscribed services to credit validation and authentication, the mobile telephone is now the individual. This, in fact, is true portability because the smart card can also carry with it the telephone number of the person it is associated with. From country to country, roaming occurs on a daily basis overseas. Yet in North America, the seamless roaming only occurs within specific carriers and networks where reciprocal agreements are in place. This means that different standards are still being used in both parts of the world.

Most of the international countries have adopted GSM as the global platform. The North American community has many standards in place; most are not compatible with each other. This is the sad truth on a recurring basis. If a user chooses to change suppliers, then the technology in the handset may have to be changed, which is not the case in the international market.

The ITU is working on a third-generation mobile standard. Unfortunately, this has been somewhat of a figment of our imagination because the standards are different in countries around the world. The North American market accounts for over 109 million wireless users, who are all moving toward a broadband communications network and the personal communications services that follow. This is a fairly significant portion of the global installations (57 percent) for wireless communications. One cannot ignore the multiple standards such as

- GSM-TDMA
- North American-TDMA (IS-54)
- CDMA (IS-95)
- TDMA (IS-136)
- AMPS
- DECT

The Need for Interoperability

Many other standards have been rolled out or implemented on a sporadic basis. The one standard that has been implemented around the world is GSM. The third-generation mobile standard can become a reality now that some settlement has occurred between many of the players (Ericsson and Qualcomm, for example), but this dream may be well into the new millennium before final realization.

Currently, no single wireless technology can claim full coverage around the world. GSM has a large footprint in Europe, Asia, and Africa. Even though GSM has also been deployed in North America, ANSI-136 (TDMA) has a large footprint in the Americas. Combined, these two technologies can claim nearly full global wireless coverage. However, both GSM and ANSI-136 have dissimilar air interface and network interface protocols. This creates an interoperability problem.

Interworking between these technologies is required to enable subscribers to roam between the TDMA-based systems, regardless of location. The *Universal Wireless Communications Consortium* (UWCC) and the North American GSM Alliance recognized these facts. In 1999, these committees formed a team to specify the desired interworking and interoperability between GSM and ANSI-136.

The *GSM / ANSI-136 Interoperability Team* (GAIT) was formed to specify a multi-mode mobile station to work across GSM and ANSI-136, and a network interworking function to translate between the two network protocols. So far, GAIT has published four technical specifications:

- GSM Hosted SMS Teleservice Specification
- GSM/ANSI-136 Common Mobile Terminal Specification
- GSM/ANSI-136 SIM Specification
- GSM/ANSI-136 Network Interworking Specification

Combining existing GSM, ANSI-136, and ANSI-41 standards, these specifications define the network elements and protocol necessary to provide basic GSM and ANSI-136 interoperability. To date, the GSM/ANSI-136 specifications outline the following:

- A multi-mode mobile station that provides operation across GSM (900, 1,800, and 1,900 MHz), ANSI-136 (800 and 1,900 MHz), and AMPS (800 MHz) systems
- A common automatic network selection algorithm for any of the systems with operator-defined priorities
- Handoff only between similar technology (GSM to GSM and ANSI-136 to ANSI-136)
- Automatic registration and de-registration
- Authentication in all modes
- Encryption in all digital modes
- SIM-based roaming in both GSM and ANSI-136 modes
- Automatic call delivery
- Transparent use of supplementary services across GSM and ANSI-136 modes, including call forwarding, call waiting, three-way calling, and messaging
- Call barring support within the capability of existing standards
- Mobile-originated and mobile-terminated short message service
- Message waiting notification
- Over-the-air activation and programming
- Circuit and packet data modes

Network Interoperability

To support mobility management and services between various *public land mobile networks* (PLMN), GSM networks use GSM *mobile application part* (MAP) messaging over SS7 signaling networks. Likewise, ANSI-136 and analog AMPS networks rely on ANSI-41 MAP messaging over SS7 to provide similar functionality. To enable subscribers to operate in both GSM and ANSI-136 networks, a network interface between GSM and ANSI-41 MAP signaling must be established. When a subscriber roams in a foreign mode (that is, GSM subscriber roaming in an ANSI-136 network or an ANSI-136 subscriber roaming in a GSM network), a conversion between native and foreign MAP protocols must be provided for service authorization and control.

To support fully automatic, two-way interoperability between GSM and ANSI-136 or AMPS, network connectivity and MAP protocol conversion are provided through a bi-directional *Interworking and Interoperability Function* (IIF). This capability was initially defined in TIA/EIA/IS-129. In terms of implementation, the IIF can be offered via a stand-alone network element or it can be integrated with the HLR.

Figure 6-7 is an overview of the expected network connectivity for interoperability. Note that the IIF may contain provisions to convert between ANSI SS7 and ITU SS7, so a separate ANSI to ITU SS7 interworking gateway is not required for all implementations.

Most users still think of mobile communications as simply a means of talking on the move. All that is about to change. Still as the interoperability takes hold, the evolution of the ANSI-136 (IS-136) standard is shown to be one that will ultimately take us to 384-Kbps data transmission as well as a voice call. Figure 6-8 is a representation of the evolution of the IS-136 standards. Work is now well advanced on the third stage of GSM standard development. Known in the business as Phase 2+, the new technology will enable GSM to serve as a platform for a whole range of sophisticated services that will keep GSM operators ahead of the game.

Enhanced Voice Services

Cellular office systems, indoor wireless systems, and microcellular systems have been the focus of a number of technology and market trials in the past. These trials have generally shown promising results for in-building

Figure 6-7
The network interoperability of GSM and ANSI-136

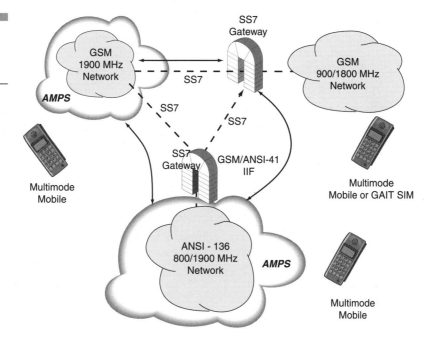

Figure 6-8
The IS-136 evolution to 3G services

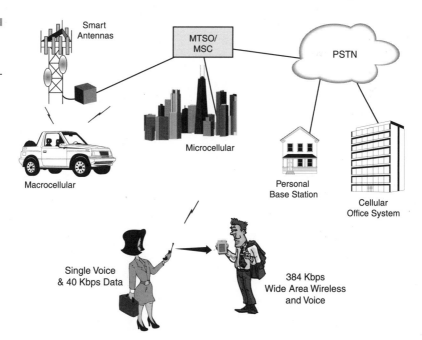

microcellular systems, particularly when the in-building systems were integrated with conventional macrocellular systems to provide seamless anywhere access. The increased ability for subscribers with a single-handset/single-number capability results in increased productivity for the subscriber and higher satisfaction with their wireless services. Furthermore, the conclusions are that users have demonstrated a willingness to pay for indoor wireless access within reason. The disadvantage is that separate spectrum is required to support these dual operation systems.

The services include features and functions such as

- More access and use of SMS: Five billion SMS text messages were sent over the world's GSM wireless networks during December 2000. The number indicates a fivefold increase in the volume of text messages generated every month by GSM wireless customers around the globe in one year. Figure 6-9 is a representation of the SMS delivery system.
- Increased data transmission speeds.
- High Speed Circuit Switched Data, giving a giant boost to data credibility.

Figure 6-9
The SMS delivery systems

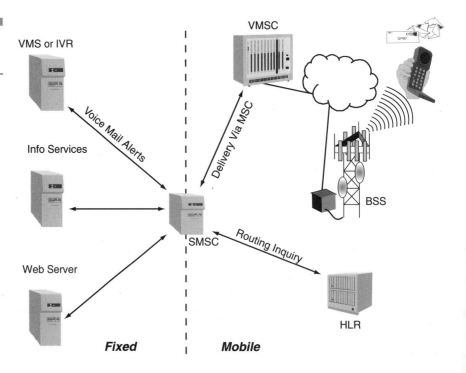

- The *General Packet Radio Services* (GPRS) standard, currently beginning its rollout, represents a shift to packet data, which will enable even more advanced services with up to 170 Kbps of packet data transmission.

- Video conferencing and high-resolution scans are just around the corner.

- GSM will facilitate Internet access, which in turn will enable business users to share information via corporate Internets, Intranets, and Extranets.

- The mobile wallet is already close to becoming a reality. Mobile phones will provide all the services currently provided by automatic teller machines and more. Customers will be able to pay bills and transfer money between checking and savings accounts just by typing in codes on their mobile terminal. Even more radical, they will be able to send electronic cash into and out of their GSM phone.

- The GSM standard is being adapted to work as a fixed access system for *Wireless Local Loop* (WLL) applications that will provide full 64-Kbps digital connections.

Add-On Technologies—iDEN™

Motorola's iDEN™ technology is a classification of SMR based on a variety of proven RF technologies using the architecture of GSM as its basis. The technology offers increased spectral efficiency and full-service integration, two of the main benefits of digital communications.

Improved Spectral Efficiency

The capacity to accommodate crowded markets and worldwide growth is a critical component of iDEN™. The development of this spectrally efficient technology enables multiple communications to occur over a single analog channel. This expansion of the network gives users greater access to the network and provides space for new and expanded services to be added without rebuilding the infrastructure.

iDEN™ represents a significant step toward the integration of wireless business communications systems that meet today's demands for a one-stop process. Motorola used a combination of technologies to create the increased capacities and the combination of services. Many of the enhancements and increased capacities come from Motorola's VSELP[1] vocoding technique and QAM modulation process as well as the TDMA channel splitting process.

Motorola's VSELP—Coding Signals for Efficient Transmission

The key to the expanded capacity is the reduced transmission rate needed to send information. Motorola has developed a vocoder technology that handles the process, as shown in Table 6-2.

This vocoder, known as VSELP (*Vector Sum Excited Linear Predictors*), compresses the voice signals to reduce the transmission rate needed to send information. Moreover, VSELP provides for clear voice transmission by digitizing the voice and providing high-quality audio under conditions that normally would result in a distorted analog voice. Using speech extrapolation, the VSELP decoder can repair the loss of a speech segment over the radio channel. The result is less distortion and interference (for example, break-up, static, and fading) as users move toward the periphery of the coverage area, enhancing the clarity and quality of voice communications at the outskirts of a cell.

Table 6-2

VSELP Coding Process Is Straight-forward

The VSELP Coding Process
1. Compresses voice signals
2. Creates digital packets of information (voice)
3. Assigns the packets a time slot
4. Transmits the information on the iDEN™ network
5. Receives the information on the iDEN™ network

[1] VSELP is a trademark of Motorola.

QAM Modulation

Whereas the VSELP compresses the signal and reduces the transmission rate, *Quadrature Amplitude Modulation* (QAM) increases the density of the information. QAM modulation technology was specifically designed to support the digital requirements of the iDEN™ network. Motorola's unique QAM technology transmits information at a rate of 64 Kbps. No other existing modulation technology transmits as much information in a narrow band channel.

Multiplied Channel Capacity

Another essential element is the *Time Division Multiple Access* (TDMA). TDMA is a technique for dividing the wireless radio channel into multiple communication pathways. In the iDEN™ system, each 25-kHz radio channel is divided into six time slots. During transmission, voice and data are divided into packets. Each packet is assigned a time slot and transmitted over the network. At the receiving end, the packets are reassembled according to their time assignments into the original information sequence.

The Advantage of Integration

Now more than ever, users are demanding multi-function devices that are simple to use. With the iDEN™ network, users need only one telephone to access voice dispatch, two-way telephony, short message service, and future data transmission. This integration provides business users with flexible communications that enable users to access information in the most efficient and convenient way no matter where they are in the system.

SMR systems are part of a larger family of products that are classified as trunked radio systems. Trunked radio involves a combination of wired and wireless communication typically found in the emergency service operation like fire, police, and road maintenance operations.

A trunked radio system always comprises several radio channels. One channel acts as the *control channel* (CC) while the others carry traffic. The CC is used for registrations, the transmission of status messages, and for call requests. This is not unlike the cellular radio-based system using the paging and control channels in a cellular operation. The difference is that in

Figure 6-10

Control and talk
channels in a trunked
radio system

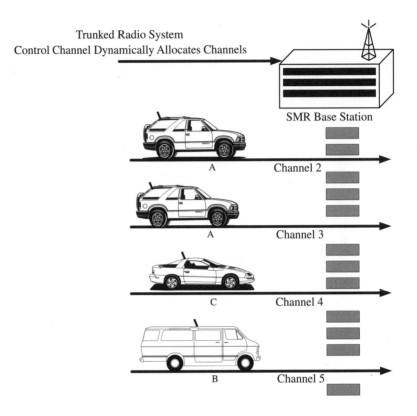

Trunked Radio System
Control Channel Dynamically Allocates Channels

SMR Base Station

A — Channel 2

A — Channel 3

C — Channel 4

B — Channel 5

a cellular radio network, several (19–21) channels are set aside, whereas in trunked radio, only one channel is needed. See Figure 6-10.

Upon requesting a call, a talk path is allocated on an exclusive basis to the subscriber from a pool of radio channels. The call is processed on this channel. If the trunked radio system receives additional call requests, a different channel is allocated to the calling party from the pool. As soon as all channels are in use, new call requests are stored in a queue. When a channel becomes available, the requested call is switched to the first available channel on a first-in first-out basis.

This method means that a call request needs to be sent only once. If the call cannot be set up immediately, the system stores the call request and processes it later.

The Control Channel (CC)

Each radio cell consists of a Trunked Radio Exchange and a *Radio Base Station* (RBS). The Trunked Radio Exchange can be used as a *Master System Controller* (MSC) or as a *Trunking System Controller* (TSC). The Trunked Radio Exchange manages the radio channels of the RBS(s). One of these channels is used as the CC. The SMR base is shown in Figure 6-11.

When a mobile set is powered on, it automatically registers using the CC. Once the subscriber receives a positive acknowledgement from the network, the mobile is registered on the trunked radio system and can be used. The mobile is constantly in contact with the CC. If a call request appears, the trunked radio system checks whether the addressed subscriber is available. If he or she is not available, not registered, or engaged, this information is given to the caller. If the requested subscriber is available, the Trunked Radio Exchange using a free traffic channel sets up the call. Status messages and short data are submitted on the CC.

Trunked radio systems share a small number of radio channels among a larger number of users. The physical channels are allocated as needed to the users who are assigned logical channels. The users only hear units on

Figure 6-11
SMR base station and radio service

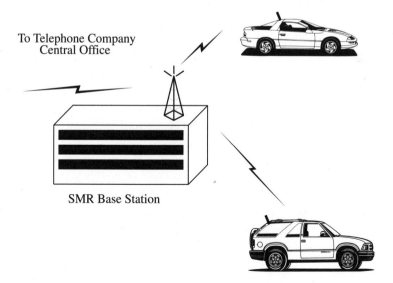

To Telephone Company
Central Office

SMR Base Station

the same logical channel. This uses the available resources more efficiently because most users do not need the channel 100 percent of the time.

Service Areas and Licensing Blocks

Two sets of frequency bands are available for SMR operation: 800 MHz and 900 MHz. Approximately 19 MHz of spectrum is available for use by SMR operators (14 MHz in the 800-MHz band and 5 MHz in the 900-MHz band). The 800-MHz SMR systems operate on two 25-kHz channels paired, whereas the 900-MHz systems operate on two 12.5-kHz channels paired. Due to the different sizes of the channel bandwidths allocated for 800 MHz and 900-MHz systems, the radio equipment used for 800-MHz SMR is not compatible with the equipment used for 900-MHz SMR.

The 900-MHz SMR service was first established in 1986 and initially employed a two-phase licensing process. In Phase I, licenses were assigned in 46 *Designated Filing Areas* (DFAs) comprised of the top 50 markets. Following Phase I, the FCC envisioned licensing facilities in areas outside these markets in Phase II. Meanwhile, licensing outside the DFA was frozen while the Commission completed the Phase I process. The freeze on licensing outside DFAs continued until 1993, when Congress reclassified most SMR licensees as *Commercial Radio Service* (CMRS) providers and established the authority to use competitive bidding to select from among mutually exclusive applicants for certain licensed services. During the freeze, however, some DFA licensees elected to become licensed for secondary sites (facilities that may not cause interference to primary licensees and must accept interference from primary licenses) outside their DFA to accommodate system expansion.

In response to Congress' reclassification of the SMR service in 1993, the Commission revised its Phase II proposals and established a broad outline for the completion of licensing in the 900-MHz SMR band. The 200 channel pairs in the 900-MHz service have been allocated in the 896- to 901-MHz and 935- to 940-MHz bands. Each MTA license gives the licensee the right to operate throughout the MTA on the designated channels, except where a co-channel incumbent licensee is operating already.

Frequencies have standard separation between the base and mobile pairs. Table 6-3 shows the operating bands for the base and mobile radio and the separation between the channels.

Table 6-3

Frequency Pairing
for SMR

Band	Base Station	Separation	Mobile Device
800 MHz	851–869 MHz	45 MHz lower	806–824 MHz
900 MHz	935–940 MHz	39 MHz lower	896–901 MHz
450–470 MHz	450–455 MHz	5 MHz higher	455–460 MHz
450–470 MHz	460–465 MHz	5 MHz higher	465–470 MHz
TV band	470–512 MHz	3 MHz higher	6-MHz TV channel

Innovation and Integration

Motorola's integrated digital enhanced network technology and protocols combine dispatch radio, full-duplex telephone interconnect, short message service, and data transmission into a single integrated business communications solution. The digital technology resulted from studies indicating that a high percentage of dispatch users carried cellular telephones and 30 percent of cellular users carry pagers, along with an increasing demand for data communications. For network design efficiency, iDEN™ uses a standard seven-cell, three-sector reuse pattern.

The technology is designed to work around many SMR spectrum limitations as well. You can take individual channels and group them together to work together as a single capacity. In cellular communications, the spectrum must be contiguous. Enhanced voice places iDEN-based services more on par with TDMA, GSM, and code division multiple access vocoders.

Although dispatch mode is simplex and not full duplex, connections are quick. It's very efficient and very fast. A typical cellular call, with speed call, would take 7 to 10 seconds for a path to be established. With Motorola's product, it takes about a second. Add in 140 character alphanumeric displays (for short message capability) and direct circuit switched data support and you get innovation and integration in one neat little package.

Spectral Efficiency with Frequency Hopping

Geotek targets the same wireless niche with its digital frequency hopping multiple access-based technology and networks. Geotek holds a near-unique position as manufacturer and spectrum owner/service operator, offering an integrated suite of mobile office solutions for dispatch, fleet management, and mobile businesses. Service offerings include dispatch, telephony, two-way messaging, automatic vehicle location, and packet data.

Frequency Hopping Multiple Access (FHMA) technology lies at the core of Geotek's networks. This TDMA and spread-spectrum derivative, originally developed by the research and development arm of the Israeli military, employs frequency hopping to achieve substantial benefits in flexibility and spectral efficiency.

FHMA achieves 25 to 30 times the capacity of existing analog technologies using a macrocell approach. Macrocells typically cover areas up to 70 miles in diameter with up to 10 radial slices or sectors, incorporating from 5 to 20 microsites.

Within sectors, FHMA implements synchronized hops from one discrete frequency to another in a predetermined manner at both the transmitter and receiver—you stay on one piece of spectrum for only a short time.

FHMA slices information packets into pieces, shuffles, and then transmits them. Packet losses result in minimal degradation because sequential losses don't occur. In addition, the system uses two-branch diversity, incorporating two separate antennas and receivers on both ends. This space diversity ensures that the best incoming signals are chosen.

The system is built on TCP/IP, and every system user has an IP address. With its inherent data integration, the Internet logically becomes a larger part in Geotek's service strategy. Geotek's automatic vehicle location capability is illustrative of the things to come. Sophisticated business data applications will drive the mobile business markets.

Digital Transition

Digital technology is an integral part of Ericsson's SMR and private radio systems, but not as an all-or-nothing requirement. Its systems can migrate to all-digital configurations, as customer needs dictate. Ericsson's enhanced digital access communications system (EDACS) technology package employs standard trunked radio.

An all-digital system, Ericsson's Aegis system uses adaptive multi-band encoding vocoder to transform analog voice signals to digital. After the vocoding process, error protection codes are added to the digitized audio stream. This process is further augmented with synthetic audio regeneration, which replaces certain portions of the voice signal corrupted with noise with usable segments of speech.

EDACS embeds its control information on the channel as well. This includes unit identification or push-to-talk identification, priority scan information, and talk group segmentation. With error correction and detection added, as well as channel signaling and synchronization, the combined signal transmits at 9,600 bits per second.

The combined voice and data encoding is an EDACS feature that not all systems can match. Many others require separate channels for voice, data, and control signaling. EDACS combines all three, automatically compensating for periods of high demand with no sacrifice in capacity or reliability.

Both the GSM and SMR (iDen) services offer significant coverage and show steady growth. In dealing with broadband communications for the future, both techniques have captured the market share necessary to compete head-on. As more providers use GSM and SMR, the industry will likely see a split in the way users select service. The two primary services will include CDMA and GSM as the future of 3G wireless communications.

Wireless Data Communications Services

Up to now, we have seen the various techniques used in a wireless communications network. This includes the general radio-based systems, the modulation techniques and the access methods, and the different forms of access such as *Time Division Multiple Access* (TDMA), *Frequency Division Multiple Access* (FDMA), *Code Division Multiple Access* (CDMA), and *Global System for Mobile* (GSM). Each of these chapters covered the benefit and the market of the systems throughout the world. Moreover, we saw the growth sector for each of the services. However, none of the topics thus far has dealt with the real issue. Instead, the coverage showed the use of a voice network strategy that the carriers have followed. And why not? The carriers went where the market was, where the revenues justified the investments and the overall demand. Voice was always the driving force behind most of the networks because it constituted more than 90 percent of the revenues derived from the networks. However, one cannot say that data would not be equally as beneficial because the networks were not built to satisfy the data communications demands. Moreover, immediately after data was introduced into the cellular and GSM markets, it became obvious that the speeds were too slow and the connections were unreliable. Without major innovation, data transmission on the wireless networks would clearly never happen. This meant that the suppliers needed some form of motivation to make the investments in data.

Over the last 25+ years, we have seen an explosion in wireless communications and computer technology. During the last 7 years, we have also seen the explosion of the Internet. The wireless data industry is at the center of this convergence. What technologies could not benefit from wireless access to the Net?

The Wireless Revolution

Twenty-five years ago, commercial uses for wireless data were largely confined to private microwave data networks used by railroad companies and specialized mobile radio systems used for dispatch services by taxi companies and the local police. Technological advances (digital radio enhancements, packet data, data compression, and smaller devices) and critical regulatory decisions (to license new spectrum for cellular telephony and other new applications) have greatly increased the availability of wireless communications while reducing costs to consumers. The result has been a dramatic growth in the market for cellular telephones. Many feel the wireless data industry is now poised for similar growth.

Voice to Data

Until recently, wireless data was also essentially a niche market largely confined to vertical applications within large companies. For example, IBM, Federal Express, and UPS built successful private wireless data networks to enable their field service personnel to operate more efficiently. The explosion of the Internet, of corporate Intranets, and the convergence of the computing and communications industries are creating new opportunities. Shortly after the interest built for data communications, we saw that wireless data could be moved in various forms, including

- Circuit-switched data over the analog cellular and digital networks
- *Cellular Digital Packet Data* (CDPD) over spare channels or dedicated channels on the existing networks
- *Short Message Services* (SMS) using the GSM standard
- Wireless Internet

Even these service offerings were not enough. Consequently, IP networking across a wireless network became the hot button. *General Packet Radio Service* (GPRS) was developed to sustain a packet data network over GSM at speeds of up to 170 Kbps. CDMA circuit-switched data at speeds of up to 64 Kbps also arrived. Protocols like the *Wireless Application Protocol* (WAP), Bluetooth, Short Message Services, and Blackberry were developed in support of wireless intelligent networking and wireless Internet. The sustained speeds and throughput are estimated at 28.8 Kbps up to 2 Mbps[1] to allow mobility and data to coexist.

The Wireless Data Market

Current revenue forecasts for the wireless data market predict strong industry growth. One major research house predicts that the *compound annual growth rate* (CAGR) for wireless data from 1996 through 2003 is projected to be 35 percent. The market is expected to grow close to $3.7 billion by the year 2004, as seen in Figure 7-1.

---

[1] Although the manufacturers claim 2 Mbps, the reality is that approximately 500 Kbps (or less) can be achieved when using a mobile device. However, if the mobile station is not moving, a higher transmission rate can be achieved over shorter distances.

Another research house projected that more than 1 million *wireless intelligent terminals* (WIT) will be sold in the year 2002 (as shown in Figure 7-2) comprising almost 4 percent of total wireless terminal sales that year.

According to a third research group, the opportunity for wireless data communication in the United States is huge, with just under 1/4 of these workers having a mobile job requirement, but growth will be slow and steady. The numbers represent 25.3 million of the 112.1 million workforce employees having a mobile job requirement (see Figure 7-3).

Figure 7-1
The wireless data
market projections

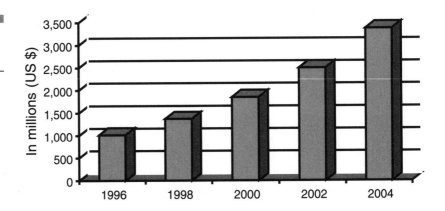

Figure 7-2
Wireless terminals in
2002 should
represent 4 percent
of total sales.

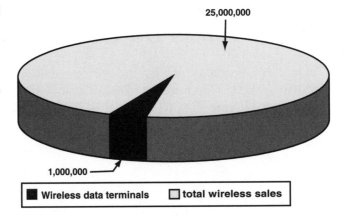

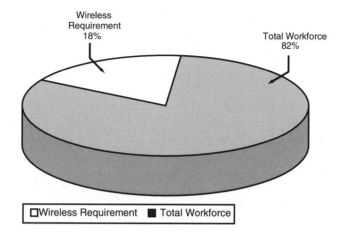

Figure 7-3
Percentage of
workforce with a
wireless data
requirement

Wireless
Requirement
18%

Total Workforce
82%

☐ Wireless Requirement ■ Total Workforce

Wireless Data and the Spectrum

The radio spectrum is the part of the electromagnetic spectrum that provides the space through which a variety of land mobile radio services operate, such as paging, cellular, private radio dispatch (taxis), microwave, television, and radio broadcasting. Two million wireless data subscribers existed in 1997 and the market growth has escalated at an average annual rate of over 40 percent, showing the number of users through 2002 as seen in Figure 7-4. This growth estimation is a relatively conservative number given the importance of the data needs today, the need for Internet capability from a mobile device, and the change in mobility of the end user. However, this 40 percent annualized growth is a good way of presenting just what has been happening for the carriers and the providers alike. This curve is one that commands respect. In 1999, for example, the number of *personal communications services* (PCS) in use grew by 22 percent, whereas the number of wireless users grew by 80 percent; nearly half of this new growth proliferated in the demand for new forms of data.

Spectrum Regulation

The *radio frequency* (RF) spectrum is a scarce and shared resource, used nationally and internationally, and subject to a wide range of regulatory oversight. In the United States, the *Federal Communications Commission*

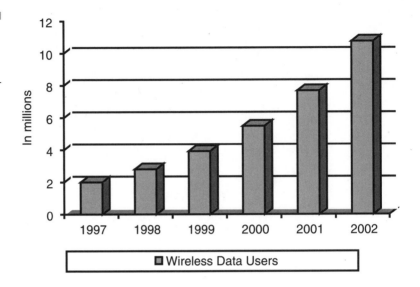

Figure 7-4
Growth in the
number of wireless
data users

(FCC) is an essential regulatory body that allocates spectrum use and resolves spectrum conflicts. The *International Telecommunication Union* (ITU) is a specialized agency of the United Nations that plays the same role internationally.

Unlicensed Spectrum

Not all portions of the spectrum are subject to licensing and regulation. Some portions are unlicensed. Three major Industrial, Scientific, and Medical bands are available for unlicensed use at 902–928 MHz (U.S. only), 2,400 MHz, and 5,800 MHz worldwide.

Wireless Data Transmission: How It Works

Until the last decade, most wireless data transmitted through radio communications was analog. Analog systems use continuous electrical signals for the transmission and reception of information. The analog systems when well designed are very strong performers and cannot be beaten. However, the analog systems are inefficient. The use of analog transmission uses the entire channel capacity (30 KHz in the U.S.) for a single voice or data call. The circuit-switched data transmission uses the standard modem that will transmit up to 28.8–33.6 Kbps across the normal channel. The analog wave (Figure 7-5) uses the continuous wave to transmit the data.

Next generation wireless data systems are turning toward the use of digital signals, whose amplitude variations with respect to time are not continuous, but discrete. They use a digital transmission service by creating the discrete values of a 0 or 1 in the representation of the square wave, as shown in Figure 7-6. The digital pulse is presented to the radio system across the radio spectrum. The transmission of digital signals using the entire channel capacity was excessive, so the use of the radio media access control led to the use of the TDMA and CDMA formats. The TDMA

Figure 7-5
The analog radio system

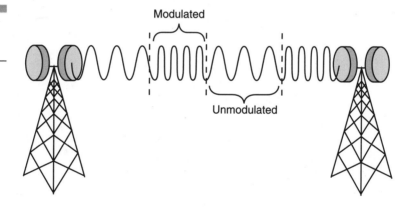

Figure 7-6
Digital radio transmission

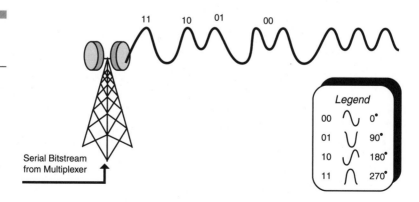

technique became accepted worldwide with the implementation of GSM in 1992. Digital *Advanced Mobile Phone Service* (AMPS) (IS-54) TDMA and now IS-136 techniques led the way in North America. Conversely, CDMA also exploded in North America. Figure 7-6 shows the modulation of a digital signal across the radio waves.

Session versus Packet Transmission

Circuit-switched (or session-based) communications assign users a discrete line or radio channel that is dedicated to the users until the session is completed. Just as two cans and one string can only handle one conversation between two people, the circuit for the data exchange is tied up until the communication is complete. Using a wireline model, modem speeds began at 300 bits per second and rapidly progressed to support 1.2, 2.4, 4.8, 9.6, 14.4, 19.2, 28.8, 33.6, and now 56 Kbps. Current state of the art speed is 33.6 or 56 Kbps. With compression, this rate can be driven to 115–230 Kbps when everything is working perfectly. Wireless data speeds are slower but improving. Table 7-1 lists the speeds of the wireless networks and modem capabilities.

The goal of a wireless data product is to replicate the following elements of the circuit-switched wireline modems:

- Functionality
- Connectivity
- Reliability
- Speed

In the users' eyes, wireless connections must perform as well as landline connections in all four categories.

- Connectivity can be defined as the capability to initially connect and stay connected through a wide variety of conditions. It can be expressed as a percentage of attempted calls. That is, if a cellular modem attempts 10 calls and only successfully connects 7 times (not including busy and no answer conditions), the connectivity would be 70 percent.

Table 7-1

Summary of Wireless Data Speeds

Wireless Networks Speeds	Current[2]	Future[3]	Mode
Motient (formerly ARDIS)	19.2/4.8 Kbps	28.8 Kbps	Packet
CDMA	64 Kbps	2.0 Mbps	Circuit/packet
W-CDMA	500 Kbps	2.0 Mbps	Circuit/packet
CDPD	19.2 Kbps	28.8 Kbps	Circuit/packet
Cellular	14.4/9.6 Kbps	28.8 Kbps	Circuit/packet
GSM	9.6 Kbps	14.4 Kbps	Circuit
GPRS/GSM	28.8 Kbps	170 Kbps	Packet
EDGE/GSM	—	384 Kbps	Packet
Metricom	33.6 Kbps	56.6 Kbps 128 Kbps	Circuit Packet
Bell South Wireless Data (BSWD) formerly RAM Mobitex	8.0 Kbps	9.6/28 .8 Kbps	Packet

[2] This speed is what is advertised; actual speeds and throughput are lower in most cases.

[3] The speeds listed are proposed for the future, but may deliver much less in a mobile environment.

■ Interoperability can be defined as the capability to operate with a variety of modems, network equipment, and conditions. It can also be expressed as a percentage of attempted calls to different modems. A number of factors affect this result including the brand of modem, protocol support, protocol implementation, network, and line conditions.

Air has combined these two concepts into a single methodology for cellular modems. This combination accurately mimics the users' experience. Users generally do not know what the cellular conditions are at a given moment and do not know what kind of modem might be on the landline side they are calling. Hence, both connectivity and interoperability become completely related and inseparable in the users' experience.

In the past, users complained about the ability to connect over a cellular modem. To solve the complaints, cellular modems were tested for performance using a cellular modem connected to a landline modem of the same brand and protocol. This also happened in a lab without actual radios in the circuit and under the best of conditions. Sure the devices could pass the connectivity test; however, under real world conditions, the results were much worse than the test results. Thus, a real world scenario is needed to make sure that the circuit-switched data communications is possible with cellular modems. Normally we need to try this under random cellular conditions, random landline conditions, random cellular locations, random landline locations, and random landline modem types. These conditions are designed to emulate the real world and must be checked. Moreover, the data throughput is another form of testing condition that must be used in this environment to see the rated speed of the modem, compared to the actual speed and throughput of the same modem under different conditions.

The randomness of the cellular conditions and locations can be achieved by testing the modem nationwide, at different times of the day. Figure 7-7 shows the typical dial-up communications session. Note that the dial-up circuit-switched networks use mostly analog connections, like the modem. Therefore, when dialing through a digital telephone set, we automatically can fall back to a dual-mode environment and pass the call across the analog portion of the circuit-switched networking standard. This figure shows the connection being made into the public-switched telephony network to access the circuit-switched CDPD, X.25, Internet, and Frame Relay and *Public-Switched Telephone Network* (PSTN) analog modem networks. The rated speed of our dial-up modems currently supports 33.6 Kbps up and up to 56 Kbps down. This depends on the circuit and the type of connection we

Figure 7-7
Wireless circuit-switched connections to other data networks

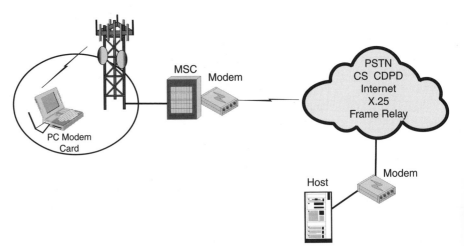

dial. Typically, we will see up to 28.8 to 31 Kbps up and 35 to 38 Kbps down on the PSTN side of the business.

Things are different, however, for wireless data transmission using a cellular modem. In fact, for the dial-up connection, the average we can expect to see is approximately 9.6 to 14.4 Kbps, assuming we have a good connection. In many cases, these are not assured speeds and may drop to a lower speed depending upon the distance from the cell site, the surrounding area, and the *signal-to-noise* (SNR) ratio. The dial-up connection shown in Figure 7-8 is a different representation of a dial-up connection through the cellular provider in a back-to-back modem pool. The modem on the cellular side is provided to accommodate the end user, whereas the standard landline modem is provided to ensure operability and connectivity in the PSTN. This connection may include modems at different speeds and services. Regardless of the modem type or the service dialed, the modem pool provides a form of connection from end-to-end into the PSTN. Once again, the speeds and throughput are conditioned upon the link, distance, and signal-to-noise ratio.

As much as these modem arrangements work, they are still too slow for most applications by today's standards. Yet, for many people, just getting a connection at 9.6 may be sufficient based on the application. If all the user has to transmit is e-mail, spreadsheet information, or a word processing

Figure 7-8
Modem pools at the carrier location

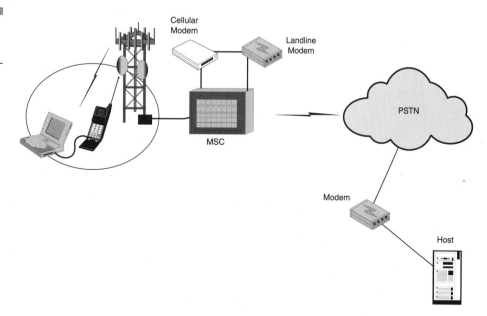

document, the speed may suffice. Where and how we define broadband communications becomes one of the challenges. To some, 28.8 Kbps of throughput may be considered "lightning speed and broadband enough," at least for today.

Where the throughput of the medium (the airwaves and PSTN combined) does not suffice, compression techniques can be employed to make sure that the speed appears greater than it actually is. In current technology, the rated equivalent throughput for a V.42bis protocol (4:1 compression) on landline services is 115.2 Kbps. However, the technology using a data speed in the air combined with the compression actually delivers anywhere from 4.8 to 21 Kbps, with overall averages of 12 to 16.8 Kbps, as seen in Figure 7-9.

These data rates are before compression, whereas V.42bis data compression can enable files to be transferred at much higher rates. For common file types (word processing, e-mail, spreadsheets, and so on), speed increases of about 3:1 to 4:1 are achieved. Conversely, some files can achieve 8:1 increases, whereas some pre-compressed file types will not achieve any benefit from compression. Thus, circuit-switched analog/digital communications has more to do before we can readily accept the throughput and speeds. With landline modem communications, the systems will still be relegated to the weakest link and current speeds for modems will still hover at 28.8 to 33.6 Kbps at most.

Figure 7-9

The overall average of throughput for compression and modem combinations

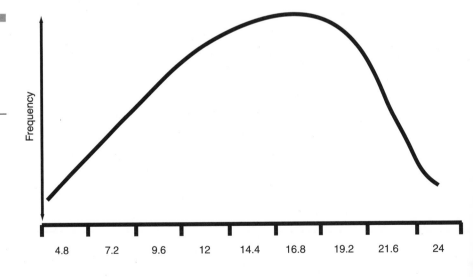

Cellular Digital Packet Data (CDPD)

IBM and a consortium of large cellular suppliers in North America developed CDPD. Most U.S. cellular operators support CDPD, however limited its use may be in the rest of the world. Having been developed as an analog transmission system, it has evolved over the years. CDPD was built with the promise of 19.2 Kbps among the North American cellular providers. It was designed to take advantage of unused space in the analog talk path, but later evolved to use unassigned voice channels to connect to Web-based services. CDPD devices search for unused voice channels using a channel-hopping approach, which allows the data to be transmitted over multiple available channels.

Circuit-Switched Cellular Digital Packet Data (CS-CDPD)

CS-CDPD enables a modem to disconnect when no data activity is present and then automatically reconnects on data flow in either direction. The short connect time makes this disconnection and reconnection transparent to the user. The inactivity timer can be adjusted so that the connection is handled in such a way that modems are not online for long periods of inactivity, thus saving end-user money. In effect, analog cellular communications can emulate a connectionless-oriented protocol like CDPD while using the connection-oriented dial-up network. Like the RAM mobile data service, CS-CDPD will automatically reconnect when a call is dropped while driving through a tunnel. The mobile data arrangement creates a hold and forward connection whereas the CS-CDPD uses a restart operation.

CDPD is an alternative way to send and receive data over the existing cellular network. The intention was to develop a method where short messages and data could be sent in between voice calls using much of the same infrastructure.

Packet-Switched CDPD

CDPD is a packet technology that sends small packets (usually up to about 1,500 bytes) of information for small bursts of time. It is an overlay on the

AMPS networks. Although technically, files of virtually any length may be sent, the network is optimized for fast, low-cost transmission of smaller files. Because the data (such as a message) is often sent in small amounts, users aren't as concerned with throughput as they would be with circuit-switched data (where you are paying for time, not data). The components function the same as networking components with which most people are already familiar. Figure 7-10 shows the CDPD naming conventions with the familiar networking names.

Comparing the CDPD layer to a common architecture such as the OSI model or the TCP/IP protocol stack helps one understand how the CDPD network operates. In Figure 7-11, the stacks are shown with standard OSI layers on the left, the TCP/IP stack in the middle, and the *local area network* (LAN) device driver at the data/network layer.

Using a mobile CDPD stack shows that few changes are needed to operate over the CDPD network. The CDPD device driver is in place of the LAN driver. This is essential to the transparency CDPD offers. The application does not need to concern itself with the fact that the network card is actually a wireless communications device.

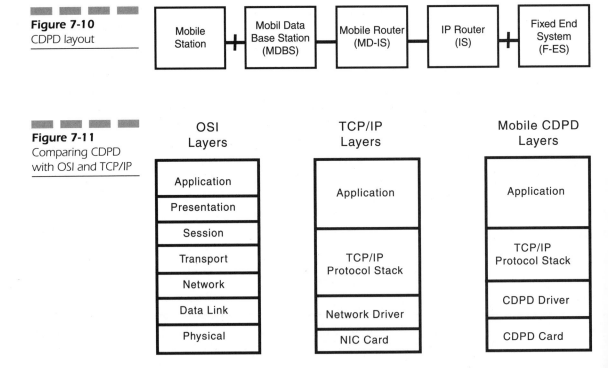

Figure 7-10
CDPD layout

| Mobile Station | Mobil Data Base Station (MDBS) | Mobile Router (MD-IS) | IP Router (IS) | Fixed End System (F-ES) |

Figure 7-11
Comparing CDPD with OSI and TCP/IP

OSI Layers	TCP/IP Layers	Mobile CDPD Layers
Application	Application	Application
Presentation		
Session		TCP/IP Protocol Stack
Transport	TCP/IP Protocol Stack	
Network		
Data Link	Network Driver	CDPD Driver
Physical	NIC Card	CDPD Card

CDPD is designed as an IP network. It does not use telephone numbers directly; rather, it uses addresses for everyone on the network. As such, you would send a message to an address, which could go through a gateway to your LAN and then to your desktop as another node on the network.

CDPD uses *Gaussian Minimum Shift Keying* (GMSK) to modulate the carrier in a full-duplex mode (forward and reverse channels). It also uses *Reed Solomon coding*, a forward error correction technique. Packet-switched CDPD is advertised as a 19.2-Kbps data networking system. Network and protocol requirements cause the raw throughput to operate at about 9,600 bits per second of actual user data on an unloaded system. The overhead requirements are close to 50 percent, which makes end users cringe. The 9.6-Kbps data rate is relatively constant after the CDPD user has grabbed a channel for packet transfer. Throughput may vary depending upon the following conditions:

- The network
- Implementation
- User load on the network (including voice)

Actual CDPD channel control by a single user (channels are shared by multiple users) can be as low as 10 percent. Thus, true user throughput in actual use with multiple users can range from around 960 bps to 9.6 Kbps depending on system load. This is equivalent to about 100 characters per second at the lowest throughput and 1,000 characters per second at the top end, excluding any compression technique that may be used. Table 7-2 compares the packet network to the conventional dial-up wireless data network. This is a good way to view the differences in the overall scheme of the data networking standards.

Packet Data Communications Are More Efficient

As just discussed in CDPD, packets from a number of different conversations or data messages can traverse the same channel. Packets are mixed on the channel, but are re-assembled correctly at the receiving end. This interleaving enables the increased capacities to serve the mass market, but also provides the broadband communications as needed. With a GPRS network, the user can link eight TDMA time slots on the uplink and eight TDMA time slots on the downlink to create a speed of 170 Kbps+ packet data transmission. This requires Phase 2 of GPRS; in Phase 1, the mobile

Table 7-2

Comparison of
CDPD and Circuit-
Switched Data

Circuit Cellular	CDPD
Same as the standard landline telephone system and modem using dial-up connections.	Packet data system designed for short bursty messages. System is optimized for wireless messaging.
Dial virtually any modem in the world and connect using conventional telephone numbers.	Cannot dial phone numbers directly, but can send messages to addresses such as IP, X.25, etc.
Complete faxing capability to any fax machine or fax modem based on conventional telephone standards.	Faxes limited to text only (through a fax service gateway). Cannot send graphical images from fax.
Cost-effective for large files and data-intense applications.	Very cost-effective for large numbers of short messages.
Expensive for large numbers of very short messages.	Expensive for e-mail attachments, faxes, large files, high data content.
Average throughput ranges from 500 to 1,500 characters per second (and 4:1 with compression).	Average throughput ranges from 100 to 1,000 characters per second (and 4:1 with compression).

can only use three time slots. This, of course, depends upon the distance from the GSM cell, the interference in the area, and the signal-to-noise ratios being achieved. The use of a packet transmission generally uses the *Internet Protocol* (IP) packets (also called datagrams) to achieve the data throughput. Some forms of packet transmission (such as CDPD) were designed to take advantage of the dead space in the time slots on an analog voice channel. Packets of data were interleaved onto the circuit when silence was on the channel. However, because the carriers moved to a digital networking standard, such as TDMA, no real dead space occurs in the time slot. Because we share the channel (30 KHz in North America and 200 KHz in Europe GSM) among several users, the slots are full. TDMA-136 in North America shares the 30 KHz channel among six users, whereas GSM shares 200 KHz channels among eight users. With the data fill being so dense, little gets in the way of spare capacity, so the CDPD and other packet data networks require a dedicated channel. Now the industry is moving to a hybrid where some channels are allocated to voice (GSM), some channels are dedicated to data (GPRS), and other channels are shared between the two services. Today, GPRS is merely an overlay on the GSM networks.

Wireless Application Protocol (WAP)

WAP is an attempt to define a standard for how content from the Internet is filtered and provided to mobile users. WAP was developed to access content from the Internet easily from a mobile terminal. The mobile industry is excited about WAP because it combines two of the fastest growing segments: wireless and the Internet. WAP is seen as a comprehensive and scaleable protocol designed for use with

- Any mobile phone (single-line display to a smart phone)
- Current or planned wireless service such as SMS, data, *unstructured supplementary services data* (USSD), and GPRS
- A variety of mobile network standards including CDMA, GSM, and *Universal Mobile Telephone System* (UMTS)
- Multiple input terminals (PDAs, keypads, keyboards, and touch-screens)

WAP incorporates a relatively simple micro-browser into the mobile phone. WAP is aimed at turning a mass-market mobile phone into a network-based smart phone.

WAP in its design was not perceived as a means of achieving speed as opposed to being utilitarian. It has been designed to

- Provide a user interface that is optimized for the small screen of a mobile handset
- Adjust to the speed and latency restrictions of mobile networks

The WAP-enabled handset accesses Web-based content and applications through a WAP gateway. WAP is bearer-independent and can be used on any digital mobile network, packet-switched mobile data network, and *third-generation network* (3G). Many operators are currently using the circuit-switched network as a bearer for WAP services. The user experience is not as robust on a circuit-switched network because of slow speeds and lengthy connection times. This means that operators must keep their service offerings simple. The advent of packet-switched networks will enhance the user experience in terms of speed. However, this will not eliminate the central issue; applications will only be compelling if they take into account the limitations of the small screen and play to the strengths of mobility. The gateway does not have to sit inside the mobile network. Any provider of

WAP services, such as content providers, *application service providers* (ASP), and virtual network operators, can invest in a gateway. This provides them with greater control over the security aspect of offering WAP services and greater control over customer data. Figure 7-12 illustrates the basic outline of a WAP server. The gateway and the session are the critical components shown in this figure, along with the data protocols.

The WAP has been touted as the mobile surfers' dream come true in the delivery of Internet access from anywhere. Surfing with a WAP-enabled mobile handset is not a great experience for the user, and applications are limited to simple information services. For example, when using the WAP-enabled telephone set and accessing an e-mail server, attachments cannot be displayed. Moreover, with all the flashy graphics and animations of the Web sites, WAP has some problems displaying the service on the handheld screen. More compelling applications that exploit the strengths of mobility and the Web are needed, but these are not yet available.

WAP alone does not provide the capability that content providers need to add a mobility value proposition to location-dependent information. Location services will become available in the networks and through global positioning technologies, but they will require more time to mature and become a mass-produced service. The pricing model is still not acceptable to most consumers. The overall trend is that these location services will become more readily available in the 2003–2004 timeframe.

Push capabilities are not yet available over WAP whereby content cannot be streamed ad hoc to a user due to the location need and the ability to

Figure 7-12
A WAP session
(Source: Ovum)

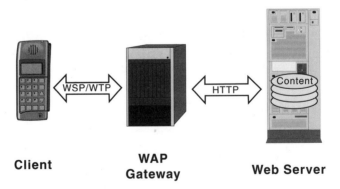

Client   **WAP Gateway**   **Web Server**

Key:
HTTP= Hypertext Transfer Protocol
WTP= Wireless Transfer Protocol

stream the form of data. Short Message Services (SMS) will be used to complement WAP in the short term, providing a pseudo-push service. However, one must be aware that not all SMS services are interoperable, particularly when the operators upgrade to accommodate other needs (such as, GPRS, EDGE, and UMTS).

In another area, the use of WAP does not assure end-to-end security which may discourage some financial service providers and other e-commerce merchants from WAP-enabling their services. Other security solutions are currently available in the form of encryption and firewalls, but this adds an element of complexity that the end user does not want to deal with.

In a *voice over Internet Protocol* (VoIP) platform, WAP will add some functionality. This will help to keep the end user interested in the use of the applications on a wireless network while waiting for newer applications and handsets to emerge. Voice calls will be set up from data calls.

SMS

The Short Message Service (SMS) has the capability to send and receive text messages to and from mobile telephones. The text can be words, numbers, or a combination of the two. SMS was created as part of the GSM Phase 1 standard. The first short message was sent in December 1992 from a *personal computer* (PC) to a mobile phone in the U.K. Each short message is up to 160 characters in length when Latin alphabets are used and 70 characters in length when non-Latin alphabets such as Arabic and Chinese are used.

There is little doubt regarding SMS success—the market in Europe alone has reached over one billion messages despite little proactive marketing by network operators and phone manufacturers. Key market drivers over the next 2 years, such as the Wireless Application Protocol (WAP), will continue this growth path.

National SMS Interworking

The addition of interworking functions between competing network operators in the same geographical market gives customers the opportunity to use SMS in the same way as they make a phone call. Just as they make a voice call to each other's phones, they can send short messages to each other. This capability rapidly increases the number of available messaging destinations, thereby increasing the value and use of SMS. As such, adding national SMS interworking can lead to a growth rate of 50 percent in SMS message volumes.

Now, the use of SMS on the network has reached critical mass. Enough regular users and momentum are behind the services. SMS has become an integral and important part of many customers' everyday business lives. Facilitating international SMS roaming was also an important development in the proliferation of SMS in countries where border crossing is frequent.

The introduction of standardized protocols such as the *Subscriber Identity Module* (SIM) Application Toolkit and the WAP contributes to an increase in messaging usage by providing a standard service development and deployment environment for application developers and business partners. These protocols also make it easier for users to reply to and otherwise access messaging services through the provision of custom menus on the phone. As such, although these protocols are only a means to an end and not new messaging destinations or services in their own right, they are likely to lead to a 10 to 15 percent uplift in total SMS volumes.

Person-to-Person Messaging

Mobile phone users routinely use the SMS to communicate with each other. Typically, this person-to-person messaging is used for the following reasons:

- Say hello.
- Remind someone of something.
- Arrange a meeting.
- Discuss something.

Such messages are usually originated from the mobile phone keypad. When the information to be communicated is short, SMS is an ideal messaging medium. For example, network operators typically charge the same price to send a short message to someone in the same room as they do to someone traveling abroad with their mobile phone. Because short messages are proactively delivered to mobile phones that are typically in the user's pocket and can be stored for later reference, SMS is often more convenient than e-mail or dial-up data. Once users become familiar with using short messages, they often find that SMS is a useful way of exchanging information and keeping in touch with friends.

Voice and Fax Mail Notifications

The most common use of SMS is to notify mobile phone users that they have new voice or fax mail messages. This is the starting point for many mobile phone users who use SMS. An alert by SMS informs the user whenever a new message is dispatched into the mailbox. Because SMS is already routinely used to alert users of new voice mail messages, this application is and will remain one of the largest generators of short messages.

Internet E-mail Alerts

Upon receiving a new e-mail, Internet users are not normally notified. They have to dial in periodically to check their mailbox. However, by linking Internet e-mail with SMS, users can be notified whenever a new e-mail is received. The e-mail alert is provided in the form of a short message that typically lists the header and the first sentence of the message. Most e-mail solutions incorporate filtering, so users are only notified of certain messages. It can be expensive or inconvenient to be notified regarding all e-mail messages (including spam). Because of the increasing usage of Internet e-mail for global communications, this application is becoming popular for SMS.

The Wireless Internet

Most of the wireless carriers are uniquely positioned to become wireless multimedia service providers, offering voice and data communication services to mobile users in this growing market. However, these carriers must understand and deploy the architecture needed. Furthermore, they must understand IPs and what it takes to support these strategic services. As wireless data proliferates and the volume of IP traffic grows, current wireless networks require a strategy similar to the wireline networks. They must take a position of being

- Reliable
- Scalable
- Manageable
- IP enabled and optimized

The *wireless Internet* (WI) will evolve as the standard for mobile IP services. The WI uses a modified and reinforced packet-based core network infrastructure optimized for IP. The current trends toward an integrated and converged network will become the reality for the future in the evolution of common packet architecture for voice and data communications.

- Voice will be data and data will be too! Voice packets will be carried as the primary network service. As data traffic increases, a reduced need for the number of physical network elements and links will occur. Moreover, the carriers will see a reduction in overall facilities costs, as the manufacturers have been promising since 1996. The benefit of using combined voice and data services may create an 8- to 11-fold increase in resource efficiency. This means that the carriers will expand their customer base without major network component investments.

- Mobility integration: services including voice, data, *mobility management* (MM), SMS, and new IP services will migrate to network servers. The reduction in total network costs and quicker time to market will benefit the carriers and the end users alike. This WI will reduce the carrier's network costs for voice to fractions of a cent per minute. At the same time, data network optimization will occur end to end. Voice and data will converge into a single network, producing better network cost benefit ratios. The evolution to a WI will expand the opportunities to introduce new applications and services due to the efficiency gains in the network.

The key developments that will drive this WI include the evolution of the following services:

- Broadband radio systems
- IP network backbones for the *public land mobile network* (PLMN)
- Wireless media access gateways and wireless gatekeepers
- IP servers
- Voice- and data-enabled Internet customer care centers

The evolution and deployment of wideband and broadband radio technologies will relieve the contention and congestion while opening the door to a readily acceptable WI. Specific radio developments include

- GPRS overlays onto the GSM network that will deliver data transmission speeds of up to 170 Kbps.

- *Enhanced Data for GSM Evolution* (EDGE) will deliver wireless data rates starting at traditional landline speeds of 56 Kbps reaching up to 384 Kbps.
- CDMA networks evolve to 1xRTT enabling data rates of 64 to 144 Kbps.
- North American IS-136 TDMA networks evolve to GPRS-EDGE; IP-based radios will deliver up to 384 Kbps.
- 3G radio standards including Global CDMA and UMTS will improve overall performance and voice density while providing data rates of up to 2 Mbps. When this is perfected, these speeds may be available to the mobile operator and end user.

General Packet Radio Systems

The impressive growth of cellular mobile telephony as well as the number of Internet users promises an exciting potential for a market that combines both innovations: cellular wireless data services. Over the past decade, an extensive demand for wireless data services has appeared. The future is no different; we will see an even larger growth curve for wireless data services. In particular, users will request high-performance WI access. As we have seen already, the in-place cellular data services do not fulfill the needs of users and providers. From the user's point of view, data rates are too slow; the connection setup takes too long and is rather complicated. Moreover, the service is too expensive for most users. From the technical standpoint, today's wireless data services are based on circuit-switched radio transmission, which is a major shortcoming. A complete traffic channel is allocated for a single user for the entire call period. If bursty traffic is present (for example, Internet traffic), this results in highly inefficient radio resource usage.

For bursty traffic, packet-switched bearer services produce better results on the traffic channels. A channel is allocated only when needed and will be released immediately after the transmission of the packets. With this principle, multiple users can share one physical channel by using a statistical multiplexing technique.

In order to address these inefficiencies, two cellular packet data technologies have been developed so far:

- CDPD (for AMPS, IS-95, and IS-136)
- GPRS

Originally, GPRS was developed for GSM. In the future, it will be integrated within IS-136. GPRS is a new bearer service for GSM that greatly improves and simplifies wireless access to *packet data networks* (PDNs). It applies a packet radio principle to transfer user data packets in an efficient way between GSM mobile stations and external PDNs. Packets can be directly routed from the GPRS mobile stations to the packet-switched data networks. Networks based on the IP and X.25 networks are supported in the current version of GPRS. Users of GPRS benefit from shorter access times and higher data rates.

In conventional dial-up GSM, the connection takes several seconds and rates for data transmission are restricted to 9.6 Kbps. GPRS offers subsecond call setup times and ISDN-like data rates. In addition, GPRS packet transmission offers a user-friendlier billing than that offered by circuit-switched services. In circuit-switched services, billing is based on the duration of the connection. This is unsuitable for applications with bursty traffic. The user must pay for the entire airtime, even for idle periods when no packets are sent (such as when the user reads a Web page). In contrast to this, with packet-switched services, billing can be based on the amount of transmitted data. The advantage for the user is that he or she can be online over a long period but will be billed based on the transmitted data volume.

GPRS improves the utilization of the radio resources, offers volume-based billing, higher transfer rates, shorter access times, and simplifies the access to PDNs. The *European Telecommunications Standards Institute* (ETSI) has standardized GPRS during the last 5 years. Currently, many pilots are being conducted and commercial rollout of Phase 1 is underway. Phase 2 of GPRS will follow right behind Phase 1 because the enhancements are in strong demand.

GPRS System Architecture

To integrate GPRS into the existing GSM architecture, a new class of network nodes, called *GPRS support nodes* (GSN), was introduced. GSNs are responsible for the delivery and routing of data packets between the mobile stations and the external PDNs.

A *serving GPRS support node* (SGSN) is responsible for the delivery of data packets from and to the mobile stations within its service area. Its tasks include packet routing and transfer, mobility management (attach/detach and location management), logical link management, and authentication and charging functions. The location register of the SGSN stores location information (current cell, current *Visitor Location Register*

[VLR]) and user profiles (*International Mobile System Identifiers* [IMSI], addresses used in the packet data network) of all GPRS users registered with this SGSN.

A *gateway GPRS support node* (GGSN) acts as an interface between the GPRS backbone network and the external packet data networks. It converts the GPRS packets coming from the SGSN into the appropriate *packet data protocol* (PDP) format (for example, IP or X.25) and sends them out on the corresponding PDN. In the other direction, PDP addresses of incoming data packets are converted to the GSM address of the destination user. The re-addressed packets are sent to the responsible SGSN. For this purpose, the GGSN stores the current SGSN address of the user and his or her profile in its location register. The GGSN also performs authentication and charging functions.

In general, a many-to-many relationship exists between the SGSNs and GGSNs.

- A GGSN is the interface to external packet data networks for several SGSNs.

- A SGSN may route its packets over different GGSNs to reach different PDNs.

Figure 7-13 shows the interfaces between the new network nodes and the GSM network as defined by ETSI. All GSNs are connected via an IP-based GPRS backbone network. Within this backbone, the GSNs encapsulate the PDN packets and transmit (tunnel) them using the *GPRS Tunneling Protocol* (GTP).

Currently two kinds of backbone networks exist:

- Intra-PLMN backbone networks connect GSNs of the same PLMN and are therefore private IP-based networks of the GPRS network provider.

- Inter-PLMN backbone networks connect GSNs of different PLMNs. A roaming agreement between two GPRS network providers is necessary to install such a backbone.

Figure 7-14 shows two intra-PLMN backbone networks of different PLMNs connected with an inter-PLMN backbone. The gateways between the PLMNs and the external inter-PLMN backbone are called border gateways. Among other things, they perform security functions to protect the private intra-PLMN backbones against unauthorized users and attacks.

The G_n and G_p interfaces are also defined between two SGSNs. This enables the SGSNs to exchange user profiles when a mobile station moves

Figure 7-13
GPRS interfaces

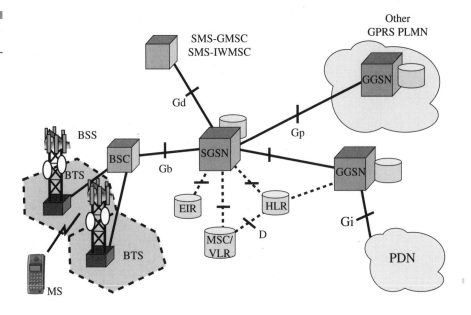

Figure 7-14
Two PLMNs
connected

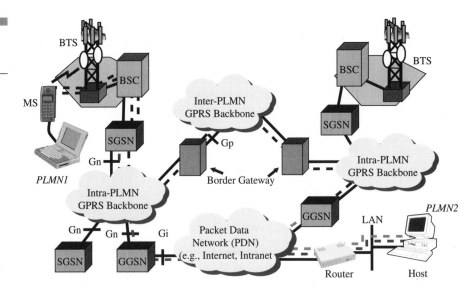

from one SGSN area to another. Across the G_f interface, the SGSN may query the *Internet Mobile Equipment Identity* (IMEI) of a mobile station trying to register with the network. The G_i interface connects the PLMN with external public or private PDNs, such as the Internet or corporate

Intranets. Interfaces to IP (IPv4 and IPv6) and X.25 networks are supported.

The *Home Location Register* (HLR) stores the user profile, the current SGSN address, and the PDP addresses for each GPRS user in the PLMN. The G_r interface is used to exchange this information between HLR and SGSN. For example, the SGSN informs the HLR about the current location of the MS. When the MS registers with a new SGSN, the HLR will send the user profile to the new SGSN. The signaling path between GGSN and HLR (G_c interface) may be used by the GGSN to query a user's location and profile in order to update its location register.

In addition, the *Mobile Switching Center* (MSC)/VLR may be extended with functions and register entries that enable efficient coordination between packet-switched (GPRS) and circuit-switched (conventional GSM) services. Examples of this are combined GPRS and non-GPRS location updates and combined attachment procedures. Moreover, paging requests of circuit-switched GSM calls can be performed via the SGSN. For this purpose, the G_s interface connects the databases of SGSN and MSC/VLR. To exchange messages of the SMS via GPRS, the G_d interface is defined. It interconnects the *SMS gateway MSC* (SMS-GMSC) with the SGSN.

Bearer Services and Supplementary Services

The bearer services of GPRS offer end-to-end packet-switched data transfer. Two different kinds are available: the *point-to-point* (PTP) service and the *point-to-multipoint* (PTM) service. The latter will be available in future releases of GPRS.

The PTP service offers the transfer of data packets between two users. It is offered in both connectionless mode (PTP connectionless network service for IP) and connection-oriented mode (PTP connection-oriented network service for X.25).

The PTM service offers the transfer of data packets from one user to multiple users. Two kinds of PTM services are available:

- Using the *multicast service* (PTM-M), data packets are broadcast in a certain geographical area. A group identifier indicates whether the packets are intended for all users or for a group of users.

- Using the *group call service* (PTM-G), data packets are addressed to a group of users (PTM group) and are sent out in geographical areas where the group members are currently located.

It is also possible to send SMS messages over GPRS.

Simultaneous Usage of Packet-Switched and Circuit-Switched Services

In a GSM/GPRS network, conventional circuit-switched services (voice, data, and SMS) and GPRS services can be used in parallel. Three classes of mobile stations are defined:

- A class A mobile station supports simultaneous operation of GPRS and conventional GSM services.

- A class B mobile station is able to register with the network for both GPRS and conventional GSM services simultaneously. In contrast to a class A mobile station, it can only use one of the two services at a given time.

- A class C mobile station can attach for either GPRS or conventional GSM services. Simultaneous registration (and usage) is not possible. Exceptions to this rule are SMS messages, which can be received and sent at any time.

EDGE—The Next Step in Wireless Data

The next generation of data heading toward third generation and personal multimedia environments builds on GPRS and is known as Enhanced Data rate for GSM Evolution (EDGE). It will enable GSM operators to use existing GSM radio bands to offer wireless multimedia IP-based services and applications at theoretical maximum speeds of 384 Kbps with a bit-rate of 48 Kbps per time slot and up to 69.2 Kbps per time slot in good radio conditions.

Vendors say that implementing EDGE will be relatively painless and will require relatively small changes to network hardware and software as it uses the same TDMA frame structure, logic channel, and 200-KHz carrier bandwidth as today's GSM networks.

As EDGE progresses to coexistence with 3G *Wideband CDMA* (W-CDMA), data rates of up to ATM-like speeds of 2 Mbps could be available. Groups from the two camps have been working on ways to converge their 3G plans, with the result that operators using either standard can roll out GPRS packet-based high-speed networks, together with EDGE as a radio interface.

GERAN

GERAN is a term used to describe a GSM- and EDGE-based 200-KHz radio access network. The GERAN is based on GSM/EDGE Release 99, and covers all new features for GSM Release 2000 and subsequent releases, with full backward compatibility to previous releases. EDGE is an evolution of GPRS that will allow up to three times higher throughput compared to GSM, using the same bandwidth. EDGE, in combination with GPRS, will deliver data services up to 384 Kbps in the near future in specific areas. As part of the overall movement to higher speed, data operators look forward to future-ready products such as radios that enable maximum re-use of an installed base, a money-saving strategy.

Although EDGE is a little further away on the horizon, it is an evolving technology that will deliver even faster throughput for both packet and circuit-switched services. EDGE redefines the GSM modulation and coding scheme that uses the same bandwidth as traditional GSM, but delivers three times the throughput. This allows end user data rates of a maximum of 48 Kbps per time slot for IP-based traffic. By adding the multiple time slot capabilities of GPRS, whereby a user can control eight time slots, data throughput rates of 384 Kbps are possible. GPRS delivers up to 10 times the speed of circuit-switched access to the Internet, and EDGE builds upon that to deliver another three times the capacity. This makes EDGE a highly cost-effective solution for operators to offer medium-speed mobile data services. Many operators and manufacturers are planning their deployments for late 2001 and commercial availability in 2002. At the same time, other manufacturers decided to abandon plans for EDGE because it was too little, too late. Instead they decided to plan for the work that was going on in parallel.

UMTS

Universal Mobile Telecommunications System (UMTS) services will launch commercially from 2001 with widespread global deployments by 2005. UMTS experimental systems are now in field trial with leading vendors worldwide. UMTS will be the next standard for mobile services across Europe: it is the European member of the IMT-2000 family of 3G cellular standards. The key benefits of UMTS include improvements in quality and services, incorporated broadband and sophisticated multimedia services, flexibility of future service creation and introduction, and ubiquitous service portability.

3G networks become a reality when they meet the UMTS standards and offer true ubiquity of an IP packet-switched backbone that can deliver any communication service anywhere. UMTS is also based on the selected Wideband CDMA (W-CDMA) technologies.

UMTS Access Network (UTRAN)

The UTRAN, based on ATM technology, will support *Frequency Division Duplex* (FDD) and *Time Division Duplex* (TDD) mode radio interfaces providing flexible, high-bandwidth bearer support, together with efficient spectrum use. UMTS base stations will provide data rates up to 384 Kbps for full mobility and up to 2 Mbps for local mobility. The UTRAN will be connected to the IP-optimized core network through gateways to provide full transparency across the backbone. Figure 7-15 shows the architecture of UMTS.

Work continues on the UMTS front. Also referred to as 3G mobile, UMTS will become the global standard for mobile wideband multimedia services and offer new business opportunities for existing and new wireless operators. Delivering data rates of up to 2 Mbps, it provides applications such as

- High-speed Web browsing
- Interactive shopping

Figure 7-15
The UMTS
architecture

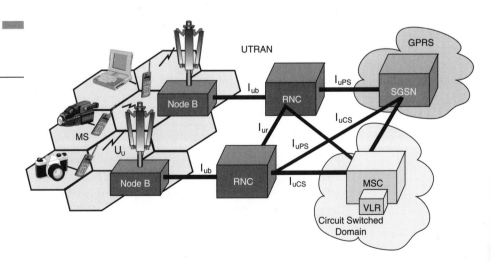

- Presentation of video clips
- Slow-scan video conferencing

UMTS will be a key component that enables the true convergence of the telecommunications, computing, and multimedia in the mobile environment. Expectations are that UMTS networks will coexist with GSM networks that have been enhanced through the application of GPRS and/or EDGE technologies.

The promise of mobile data will become a reality in the form of IP services over wireless when the industry succeeds in developing compelling applications for consumers and providing customized services that are reliable and easy-to-use. Various activities are underway to develop solutions for mobile Internet applications.

Some approaches are based on the IP server architecture, which is technology neutral. It will use software that operates on any major computer vendor's hardware. It provides a uniform interface that couples the Web, Internet, and Intranet standards, and can support manufacturer-independent terminal devices. In the past, the manufacturers' proprietary solutions have been a nuisance and a limitation to the future of wireless data communications. Now this transparency will use an OSI model to convert and format any data into a service that can be delivered to any kind of terminal, regardless of its technology including SMS phone, pager, or the WAP-enabled terminal. New services will enable wireless operators to offer seamless access to the following:

- Data services
- General information
- Stock quotes
- Weather forecasts
- Traffic reports
- Location-related services
- E-commerce
- Reservation and ticketing services
- E-mail delivery to a user's various terminals

The explosive and universal acceptance of the Internet is the catalyst for future WI and data services. Soon Web access will be as straightforward and reliable as wireline telephony. Expertise in data communications, IP, radio engineering, and switching are a given in each discipline. The new direction is to finalize the convergence revolution in the form of mobile data.

The Wireless Data Industry

Current estimates of the potential of wireless data industry range as high as $37.5 billion in revenues for the year 2002 for WI applications alone. The wireless data industry, composed of system integrators, software and hardware developers, and carriers, is defined by

- Unique devices
- Network infrastructures
- New and special services
- Internet linkages
- Specialized content

System Integrators

System integrators help pull the pieces together for clients, especially in complex vertical market applications where a single off-the-shelf solution has yet to materialize. Such applications often involve getting old devices to do new things or linking company databases for specialized field service applications. The box will be the equivalent of the Microsoft Windows™ solution for the wireless world. Wireless operators are uniquely positioned to become wireless multimedia service providers, offering voice and data communication services to mobile users in this growing market.

Software

Wireless data software firms include both wireless middle-ware specialists and new entrants from the computing environment now offering desktop-to-palmtop connectivity and integration. For example, although 3Com has discontinued the Audrey™, the ability to log onto the Internet and synchronize the Palm *personal digital assistant* (PDA) was a start in the direction for the future. The Audrey is reborn as a wireless interface to the Internet and uses its *Infrared Data Association* (IrDA) interface to synch with the PDA. The combinations are exciting and market driven to make such a device more palatable.

Hardware Developers

Building the right device, called the "unconscious carry" by one industry expert, is a technical and creative challenge. How much memory can be crammed into a device that fits in a pocket? How many *applications program interfaces* (APIs) can be embedded into the memory of the device? The use of a thoughtless process to connect to one's data, regardless of the location, is what makes this all possible and imaginable. The hardware developers have to think beyond the box and into the wireless interconnectivity and applications-driven device mode.

Carriers

The customer doesn't care how the data gets there—they just want it to get there. The amorphous cloud that we typically define for the wireline business is now becoming the accepted norm for the wireless business. When a user wants to send data, the need is to transparently move the information from one point to another. If it becomes any more complex than that, the user will balk at using this service. The pioneers of the industry who want to be the first users accessing data at high speeds will accept a less than simple service. These users tend to be more technically astute, whereas the general consuming public only wants to know, "What is it?" "What is it going to do for me?" and "How much is it going to cost me?" Anything requiring more thought than that is going to delay the overall acceptance of the technologies. When we begin using a circuit-switched data transmission capability that is the same as a wireline connection, it becomes intuitive. When the data can be sent via a packet mode transfer in a non-dial arrangement, it becomes the norm. As the carriers roll out their new data services, they must keep the end user in mind. Charging for the data will also have to be more aligned to the way that the customer does business. Some models are being built upon the DoCoMo wireless data model in Japan. DoCoMo charges by the packet, 1/4 cent per packet of data usage. What about discarded or corrupt packets, though? How might we expect to get a more reliable network to carry the data and have fewer data errors and packet retransmissions? The model must account for this or the carriers will experience some resistance from the end-user population.

Wireless Data: Apparatus Types

No single device meets the needs of every customer; not everyone wants a wristwatch communications device for voice, as shown in Figure 7-16, and even fewer may have a need for the e-mail and data transfer via this device. For some users, small size is critical—does it fit in a pocket? For others, performance and flexibility are the most critical elements of their decision. Some of the choices may be better represented by the user perspective. However, Table 7-3 lists some of the devices that will fit the apparatus need and description.

Figure 7-16
The wristwatch model does not satisfy every user.

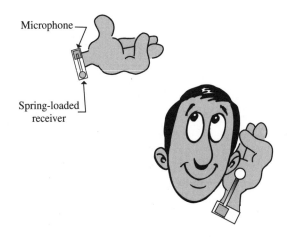

Microphone

Spring-loaded receiver

Table 7-3

Comparison of Devices That May Become Available

Item	Description
Personal digital assistant (PDA)	Palm, Visor, Newton
Palm computing device	HP and Toshiba (Libretto)
Pocket Paging device	Accompli
New appliance devices	Audrey[4]
PC cards	Airlink, Airnet-type services

[4] 3Com has discontinued Audrey, but some other device mimicking the appliance will likely appear.

However, we cannot forget that the user's degree of mobility will set the pace for the actual data throughput as already pointed out herein. When we look at the overall acceptance and the ratio of new users logging onto data networks, especially the Internet, the numbers are dramatic, as seen in Figure 7-17. The worldwide Internet users by type of access show that the growth is escalating in the wireless arena. Figure 7-18 illustrates the

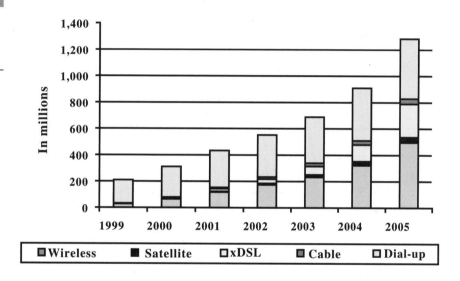

Figure 7-17
Worldwide Internet users by type of access

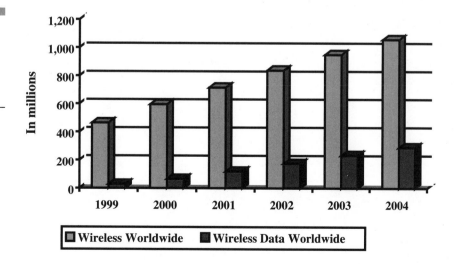

Figure 7-18
Worldwide wireless and wirless data users show a 27 percent penetration by 2004.

number of subscribers for voice only and data access. This figure also represents the growth of wireless data users on the Internet over the next few years and the ratio shift from a voice only to a data and voice application.

These figures indicate just how dependent that the world will become on wireless communications in the future. More specifically, the use of data communications will become paramount. We must wait and see what the overall strategy will be for voice and data for the future.

Wireless Local Area Networks (WLANs)

The market for wireless communications has enjoyed tremendous growth. Wireless technology now reaches or is capable of reaching virtually every location on the face of the earth. Hundreds of millions of people exchange information every day using pagers, cellular telephones, and other wireless communication products. With the tremendous success of wireless telephony and messaging services, it is hardly surprising that wireless communication is beginning to be applied to the realm of personal and business computing. No longer bound by the harnesses of wired networks, people will be able to access and share information on a global scale nearly anywhere they venture.

Wireless Local Area Networks

Looking at all of the possible combinations of a broadband wireless architecture includes finding opportunities to ramp up the speeds on the *local area network* (LAN). Mobile communications and now, the mobile data networking, are two of the areas where transmission is contingent upon the following:

- Distances from the cell
- Interference from other devices in the environment
- Power output capabilities of the mobile set
- Overall distance/speed ratio for the mobile device

Each of these factors will have a direct impact upon the speed and reliability of the data throughput in a mobile communicating service. However, the use of high-speed communications in the mobile environment, as shown in the previous chapters, was on an overall transmission of up to 2 Mbps with *Universal Mobile Telephone System* (UMTS). Yet, the speeds for mobile devices today equates to 64 to 384 Kbps, which is impressive for data transmission but not when compared to some of the wireline services. When the device is not as mobile, higher speeds may well be achieved. In a LAN, a wireless approach can be just what the data doctor ordered!

In LANs, we have seen wired speeds move up the overall curve from an Ethernet speed originally at 10 Mbps to today's current Ethernet speed at 10 Gbps. This is not ubiquitously available, but from a standards perspective, it is where the industry is heading. Figure 8-1 is a chart of the ramp up in speeds and capabilities of the Ethernet LANs over the years. Using a

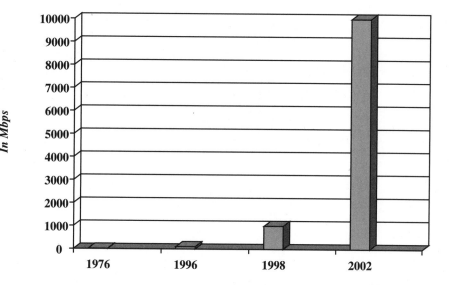

Figure 8-1
Wired Ethernet
speeds over time

fiber-based architecture, speeds of up to 10 Gbps are now achievable. But, for a wireless environment, we have seen a different approach. Instead of having unlimited bandwidth, we have been using wireless LAN technologies that deliver much less than the wired networks. Reality shows that 2 Mbps speeds were the norms of the past. However, newer techniques enable us to run at Ethernet speeds of up to 11 Mbps today. We do not actually achieve the full 11 Mbps, but then again we saw that wired Ethernet only delivered between 33 to 40 percent of its rated speeds.

Figure 8-2 is a comparison of the actual throughput of a wired LAN and a wireless LAN. The speeds are not as important as the overall throughput, but the way we measure things is where everyone gets excited. Note that the speeds of the wired and the wireless networks are impacted by more than just the medium; the protocols add a significant amount of overhead to the transmission and therefore reduce the overall throughput. One must always be aware of what is being sold as the raw data rate and what can actually be achieved when considering the use of a network topology and medium. Whereas wireline networks can achieve higher data throughput, the wireless networks provide the mobility and flexibility unavailable on a wired network. This trade-off is one of the benefits of reviewing the procedures and capabilities without the hype. Remember that there is the sizzle, and then there is the steak!

Figure 8-2
Actual vs. rated
speed comparison

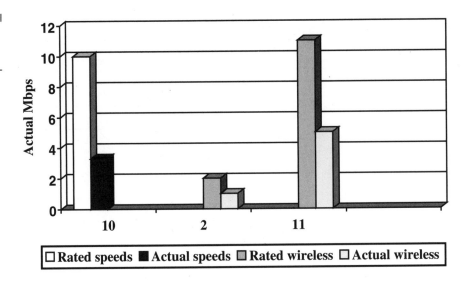

Figure 8-2
Actual vs. rated
speed comparison

Defining the Wireless LAN

A *wireless LAN* (WLAN) is a flexible data communication system implemented as an extension to, or as an alternative for, a wired LAN within a building or campus. One should consider these as complimentary products rather than competitive products.

WLANs use electromagnetic waves to transmit and receive data, such as

- Radio
- Light
- *Infrared* (IR)

WLANs transmit and receive data over the air, minimizing the need for wired connections. WLANs combine data connectivity with user mobility. Through simplified configuration possibilities, they enable movable LANs. Many of the network users are much more mobile within the office than ever before. However, combinations of LANs are available that play into the everyday installation and operation. The combinations of a LAN include a fully wired LAN, as shown in Figure 8-3. In this scenario, the LAN is installed on copper to every user. All applications and all users are interconnected across the copper-based network. The use of unshielded twisted pairs of wire make this easy to install and economical to maintain.

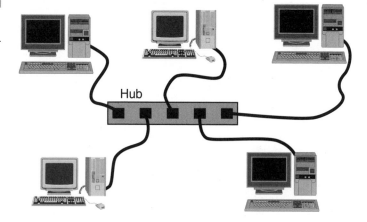

Figure 8-3
Wired LAN

Not all LANs are wired; in some cases, the network is installed as a wireless LAN based on each terminal device using a wireless interface (NIC card). The ad hoc nature of the connection is what is achieved. This form of a WLAN has always been considered the most limited because of the distances and the coverage areas in a large office building where the environment is surrounded by concrete and steel—all of which limit throughput and distances. This WLAN can be in the form of a radio-based link (Figure 8-4) or an infrared-based link (Figure 8-5). Regardless of the medium,

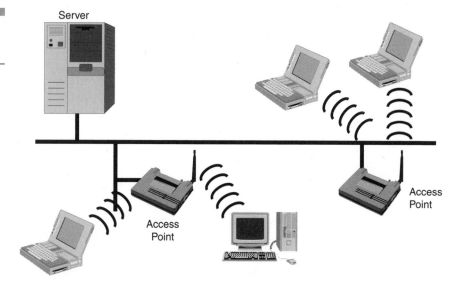

Figure 8-4
Radio-based wireless LAN

Figure 8-5
Infrared-based
wireless LAN

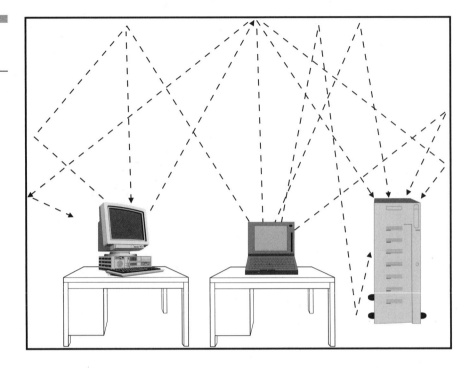

the throughput and distances are affected by the medium; the LAN in this case is very limited.

Over the last decade, WLANs have gained popularity in a number of vertical markets, including the health-care, retail, manufacturing, warehousing, and academic arenas. These industries have profited from the productivity gains of using handheld terminals and notebook computers to transmit real-time information to centralized hosts for processing. Today, WLANs are becoming more widely recognized as a general-purpose connectivity alternative for a broad range of business customers. The U.S. wireless LAN market is rapidly approaching $1 billion in revenues.

Applications for Wireless LANs

Wireless LANs frequently augment, rather than replace, wired LAN networks, providing the final few meters of connectivity between a backbone network and the in-building or on-campus mobile user. The following list describes some of the many applications made possible through the power and flexibility of wireless LANs:

- Medical staff in hospitals use handheld *personal data assistants* (PDAs) or notebook computers with WLAN capability to deliver patient information instantly.

- Consulting or accounting audit teams or small workgroups improve communications with quick or ad hoc network setup.

- Network managers in dynamic environments minimize the overhead of moves and changes.

- Corporate training sites and students at universities use wireless connectivity to facilitate access to learning information.

- Network managers installing networked computers in older buildings find that wireless LANs are a cost-effective network infrastructure solution.

- Retail store *information systems* (IS) managers use wireless networks to simplify frequent network reconfiguration and allow sidewalk and parking lot sales to occur without the heavy cost of wiring these locations.

- Trade show workers minimize setup requirements by installing pre-configured wireless LANs and reducing local MIS support.

- Warehouse workers use WLANs to exchange information with central databases within the shop.

- Network managers use WLANs to provide backup for mission-critical applications running on wired networks.

- Restaurant staff and car rental service representatives provide faster service with real-time customer information input and retrieval. The use of a handheld input *radio frequency* (RF) device prevents people from standing in long lines and speeds their service delivery.

- Managers in conference rooms make quicker decisions when they have access to real-time information at their fingertips.

- Historical building owners use wireless LANs because of limits on what can be done to the structure of the building (for example, no coring floors).

Benefits of WLANs

The widespread strategic reliance on networking among competitive businesses and the meteoric growth of the Internet and online services are strong testimonies to the benefits of shared data and shared resources.

With wireless LANs, users can access shared information without looking for a place to plug in, and network managers can set up or augment networks without installing or moving wires. Wireless LANs offer the following productivity, service, convenience, and cost advantages over traditional wired networks:

- *Mobility improves productivity and service.* Wireless LAN systems can provide LAN users with access to real-time information anywhere in their organization. This mobility supports productivity and service opportunities not possible with wired networks.

- *Wireless LANs have installation speed and simplicity.* Installing a wireless LAN system can be fast and easy and can eliminate the need to pull cable through walls and ceilings.

- *Sometimes it is more economical to use a wireless LAN.* For instance, in old buildings, the cost of asbestos cleanup or removal outweighs the cost of installing a wireless LAN solution. In other situations, such as a factory floor, it may not be feasible to run a traditional wired LAN. Wireless LANs offer the connectivity and the convenience of wired LANs without the need for expensive wiring or rewiring.

WLANs provide all the functionality of wired LANs, but without the physical constraints of the wire. WLAN configurations include independent networks (offering peer-to-peer connectivity) and infrastructure networks (supporting fully distributed data communications).

- Point-to-point local area wireless solutions, such as LAN-LAN bridging and *personal area networks* (PANs), may overlap with some WLAN applications, but they fundamentally address different user needs. A wireless LAN-LAN bridge is an alternative to cable that connects LANs in two separate buildings, as shown in Figure 8-6. A wireless PAN typically covers the few feet surrounding a user's workspace and has the capability to synchronize computers, transfer files, and gain access to local peripherals.

- Wireless LANs should not be confused with *wireless metropolitan area networks* (WMANs), packet radio often used for law-enforcement or utility applications.

- They should also not be confused with *wireless wide area networks* (WWANs), wide area data transmission over cellular or packet radio. These systems involve more costly infrastructures, provide lower data rates, and require users to pay on a time or usage basis.

On-premise WLANs require no usage fees and provide 100 to 1,000 times the data transmission rate of the exterior networks.

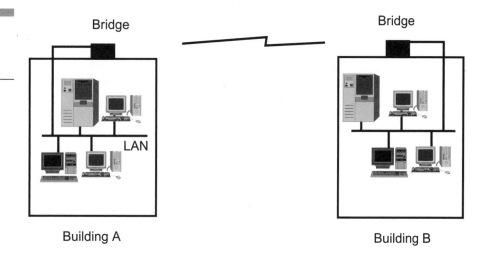

Figure 8-6
Wireless bridges connect two buildings.

How WLANs Work
===============

Wireless LANs use electromagnetic airwaves (radio and infrared) to communicate information from one point to another. Radio waves are often referred to as radio carriers because they simply perform the function of carrying the energy to a remote receiver. The data being transmitted is superimposed on the radio carrier so that it can be accurately extracted at the receiving end. This is generally referred to as modulation of the carrier by the information being transmitted. Once data is superimposed (modulated) onto the radio carrier, the radio signal occupies more than a single frequency because the frequency or bit rate of the modulating information adds to the carrier.

Multiple radio carriers can exist in the same space at the same time without interfering with each other if the radio waves are transmitted on different radio frequencies. To extract data, a radio receiver tunes in (or selects) one radio frequency while rejecting all other radio signals on different frequencies.

In a typical WLAN configuration, a transmitter/receiver (transceiver) device, called an *access point*, connects to the wired network from a fixed location using standard Ethernet cable. At a minimum, the access point receives, buffers, and transmits data between the WLAN and the wired network infrastructure. A single access point can support a small group of users (Figure 8-7) and can function within a range of less than 100 to several hundred feet. The access point (or the antenna attached to the access

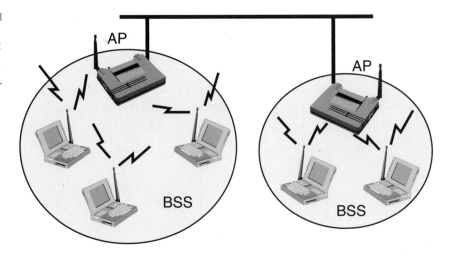

▄▄ ▄▄ ▄▄ ▄▄
Figure 8-7
Access points support
short distances and
small groups of users.

point) is usually mounted high, but it may be mounted essentially any-where practical as long as the desired radio coverage is obtained.

End users access the WLAN through wireless LAN adapters, which come in the form of

- PC cards in notebook computers
- ISA or PCI adapters in desktop computers
- Fully integrated devices within handheld computers

WLAN adapters provide an interface between the client *network operating system* (NOS) and the airwaves (via an antenna). The nature of the wireless connection is transparent to the NOS.

WLAN Configurations

Many choices are available from the simple to more complex configurations. The benefit is that these configurations can change on short notice.

Independent WLANs

The easiest WLAN configuration is probably an independent (or peer-to-peer) WLAN. Two PCs connect with wireless adapters, either RF or IR. Any time two or more wireless adapters are within range of each other, they can set up an independent network. These on-demand networks are fast and easy to use because they do not require any expensive devices or configuration. Anyone who has ever used the IR port in a PC knows how simple it can be to establish a connection with another device and exchange data. This can also be done with the IR port on a cellular phone for uploading phone lists. Figure 8-8 shows this form of WLAN.

Extended WLANs

Access points can extend the range of independent WLANs by acting as a repeater, effectively doubling the distance between wireless PCs, as shown in Figure 8-9. These access points create the effect of a hub or a switch that we typically see in a closet where a number of wired PCs are homed into a star configuration. The hub was popularized in 1984 when 10BaseT networks were introduced. By using this access point as a substitute to the hub, the WLAN can be extended beyond two independent devices. Normally the short distances between the two independent devices can be doubled with the use of the access point.

Infrastructure WLANs

In infrastructure WLANs, multiple access points link the WLAN to the wired network and enable users to efficiently share network resources. The

Figure 8-8
Independent WLAN

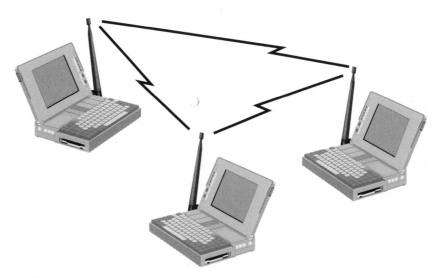

Figure 8-9
Access points extend
the WLAN distances
for wireless PCs.

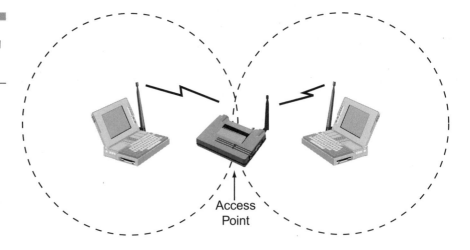

Access
Point

access points not only provide communication with the wired network, but also mediate wireless network traffic in the immediate neighborhood. Multiple access points can provide wireless coverage for an entire building or campus.

WLAN Technology Options

Manufacturers of WLANs have a range of technologies to choose from when designing a WLAN solution. Each technology comes with its own set of advantages and limitations.

Spread Spectrum

Most WLAN systems use spread spectrum technology, which has already been discussed in the *Code Division Multiple Access* (CDMA) techniques, for use in reliable, secure, and mission-critical communications systems. Spread spectrum trades off bandwidth efficiency for reliability, integrity, and security. The WLAN consumes more bandwidth than in a narrowband transmission, but the trade-off produces a signal that is stronger and easier to detect. This assumes that the receiver knows the parameters of the spread spectrum signal being broadcast. If a receiver is not tuned to the right frequency, a spread spectrum signal looks like background noise.

WLAN Customer Considerations

Compared with wired LANs, WLANs provide installation and configuration flexibility as well as the freedom that is inherent in network mobility. However, when considering the use of a WLAN, the network manager must take into account some or all of the following issues.

Range/Coverage

The distance over which RF waves can communicate is a function of product design (including transmitted power and receiver design) and the propagation path, especially in indoor environments. Interactions with objects, including walls, steel beams, file cabinets, people, doors, and so on, can affect how energy propagates. The result is that these objects affect the range and coverage of a particular system. Most WAN systems are RF based because radio waves can penetrate many indoor walls and surfaces. The range for typical WLAN systems varies from approximately 50 to 500 feet. Coverage can be extended via roaming using a microcell installation.

Throughput

As with wired LAN systems, actual throughput in wireless LANs is product and setup dependent. Factors that affect throughput include airwave congestion (number of users), propagation factors such as range and multipath, the type of WLAN system used, as well as the latency and bottlenecks on the wired portions of the WLAN. Typical data rates range from 1 to 11 Mbps. Users of traditional Ethernet LANs generally experience little difference in performance when using a WLAN. WLANs provide sufficient throughput for the most common LAN-based office applications. Heavy graphic orientation or streaming video/voice may have a significant degrading factor on the WLAN. The consideration of using a WLAN for these applications is application dependent.

Integrity and Reliability

Wireless data technologies have been used for more than 50 years. Wireless applications in both commercial and military systems are proven technologies. Although radio interference can cause degradation in throughput, it is limited in the workplace. Robust designs and the limited distance result in connections that are far more robust than cellular phone connections and provide data integrity performance equal to or better than wired networking.

802.11 Specifications

Progress and development in silicon technology coupled with lower prices and product interoperability have led to the 802.11b high-rate wireless LANs specification. These networks are now moving into the enterprise. IEEE 802.11b uses *direct sequence spread spectrum* (DSSS) technology. DSSS modulates the data string of ones and zeros with a coded chip sequence. In 802.11, that coded chip sequence is called the Barker code, which is an 11-bit pattern (10110111000). The coded chip has certain mathematical properties that lend it toward modulating radio waves. The original data is coded with the Barker code. This creates a series of chips. Each bit is encoded by the 11-bit Barker code, and each group of 11 chips encodes 1 bit of data.

Wireless radios generate a 2.4-GHz carrier wave (2.4 to 2.483 GHz) and modulate that wave using a variety of techniques. Table 8-1 is a sampling of what might be done to create the bit streams.

With 802.11 devices, as you move away from the radio, the radio adapts and uses a less complex (and slower) encoding mechanism to send data. In 1998, Lucent Technologies and Harris Semiconductor jointly proposed a standard called *Complementary Code Keying* (CCK) to the IEEE. To achieve 11 Mbps, vendors needed to change their encoding techniques. Instead of the accepted Barker code, they use codes words called *Complementary Sequences*. Using 64 unique code words to encode the signal, up to 6 bits can

Table 8-1

Samples of Modulation for 802.11

Speed	Modulation Technique Used
1-Mbps	*Binary Phase Shift Keying* (BPSK): BPSK uses one phase shift for each bit.
2-Mbps	*Quadrature Phase Shift Keying* (QPSK): QPSK uses four phases (0, 90, 180, and 270 degrees) to encode 2 bits of information. Power must be increased or distance decreased to maintain signal quality.

be represented per code word (2^6). The code word is then modulated with the QPSK technology used in 2-Mbps wireless DSSS radios. This enables an additional 2 bits of information to be encoded in each symbol. Eight chips are sent for each 6 bits, but each symbol encodes 8 bits because of the QPSK modulation. It is more difficult to demodulate the 64 code words coming across the airwaves because of the complex encoding. Radio receiver design is also significantly more difficult.

IEEE 802.11 Architectures

In IEEE's standard for wireless LANs (IEEE 802.11), a network can be configured in two different ways: ad hoc and infrastructure. In the ad hoc network, computers are brought together to form a network on the fly. The network has no structure or fixed points; usually every node is able to communicate with every other node. A good example of this is the aforementioned meeting where employees bring laptop computers together to communicate and share design or financial information. Although it seems that order would be difficult to maintain in this type of network, algorithms such as the *spokesman election algorithm* (SEA) have been designed to elect one machine as the base station (master) of the network with the others being slaves. Another algorithm in ad hoc network architectures uses a broadcast and flooding method to all other nodes to establish who's who.

The second type of network structure used in wireless LANs is the infrastructure. This architecture uses fixed network access points with which mobile nodes can communicate. These network access points are sometimes connected to landlines to widen the LAN's capability by bridging wireless nodes to other wired nodes. If service areas overlap, handoffs can occur. This structure is very similar to the present-day cellular networks around the world.

IEEE 802.11 Layers

The IEEE 802.11 standard places specifications on the parameters of both the *physical* (PHY) and *medium access control* (MAC) layers of the network. The PHY layer, which actually handles the transmission of data between nodes, can use either direct sequence spread spectrum, frequency-hopping spread spectrum, or IR pulse position modulation. IEEE 802.11 makes provisions for data rates of either 1 Mbps or 2 Mbps and calls for operation in the 2.4- to 2.4835-GHz frequency band (in the case of spread spectrum transmission), which is an unlicensed band for *industrial, scientific, and medical* (ISM) applications, and 300 to 428,000 GHz for IR transmission. Infrared is generally considered to be more secure to eavesdropping because IR transmissions require absolute line-of-sight links (no transmission is possible outside any simply connected space or around corners), as opposed to radio frequency transmissions, which can penetrate walls and be intercepted by third parties unknowingly. However, IR transmissions can be adversely affected by sunlight, and the spread spectrum protocol of 802.11 does provide some rudimentary security for typical data transfers.

The MAC layer is a set of protocols that is responsible for maintaining order in the use of a shared medium. The 802.11 standard specifies a *carrier sense multiple access with collision avoidance* (CSMA/CA) protocol. In this protocol, when a node receives a packet to be transmitted, it first listens to ensure no other node is transmitting. If the channel is clear, it then transmits the packet. Otherwise, it chooses a random back-off factor, which determines the amount of time the node must wait until it is allowed to transmit its packet. The transmitting node decrements its back-off counter during periods in which the channel is clear. (When the channel is busy, it does not decrement its back-off counter.) When the back-off counter reaches zero, the node transmits the packet. Because the probability that two nodes will choose the same back-off factor is small, collisions between packets are minimized. Collision detection, as is employed in Ethernet, cannot be used for the radio frequency transmissions of IEEE 802.11. The reason for this is that when a node is transmitting, it cannot hear any other node in the system that may be transmitting because its own signal will drown out any others arriving at the node.

Physical Signals

The wireless physical layer is split into two parts: the *Physical Layer Convergence Protocol* (PLCP) and the *Physical Medium Dependent* (PMD) sublayer.

- The PMD takes care of the wireless encoding.
- The PLCP presents a common interface for higher-level drivers to write to and provides carrier sense and *Clear Channel Assessment* (CCA), which is the signal that the MAC layer needs to determine whether the medium is currently in use. Figure 8-10 shows the PLCP.

Figure 8-10
PLCP frame for
802.11

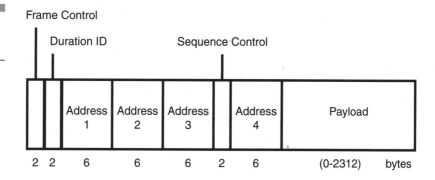

Frame Control

Duration ID

Sequence Control

| | | Address 1 | Address 2 | Address 3 | | Address 4 | Payload |

2 2   6   6   6   2   6   (0-2312)   bytes

Timing Is Everything

The most basic portion of the MAC layer is the capability to sense a quiet time on the network and then choose to transmit. Once the host has determined that the medium has been idle for a minimum time period, known as *Distributed Coordination Function* (DCF) *Inter-Frame Spacing* (DIFS), it may transmit a packet. If the medium is busy, the node must wait for a time equal to DIFS, plus a random number of slot times. The time between the end of the DIFS period and the beginning of the next frame is known as the *contention window*. Figure 8-11 is a representation of this timing window.

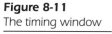

Figure 8-11
The timing window

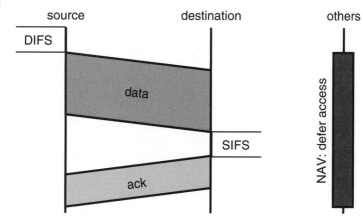

source   destination   others

DIFS

data

SIFS

ack

NAV: defer access

Each station listens to the network, and the first station to finish its allocated number of slot times begins transmitting. If any other station hears the first station talk, it stops counting down its back-off timer. When the network is idle again, it resumes the countdown. In addition to the basic back-off algorithm, 802.11 adds a back-off timer that ensures fairness. Each node starts a random back-off timer when waiting for the contention window. This timer ticks down to zero while waiting in the contention window. Each node gets a new random timer when it wants to transmit. This timer isn't reset until the node has transmitted.

Clear to Send?

In the hidden-node problem shown in Figure 8-12, workstations A, B, and C are in the same network. Workstations A and B can see one another, and B and C can see one another, but A can't see C. This happens often in real-world wireless environments, where walls and other structures create obscure radio coverage areas or when a mobile networking strategy is used. To handle this situation, a *request to send / clear to send* (RTS/CTS) is specified as an optional feature of the IEEE 802.11b standard, as shown in Figure 8-13. RTS/CTS solves the hidden-node problem in the following fashion.

When node A wants to transmit some data to node B, it first sends an RTS packet. The RTS packet includes the receiver of the data transmission ensuing and the duration of the whole transmission, including the *acknowledgement character* (ACK) related to it. Node B hears this request. Node A

Figure 8-12
The hidden-node
problem

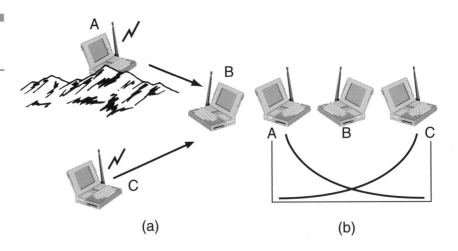

(a)   (b)

Figure 8-13
The RTS/CTS solution

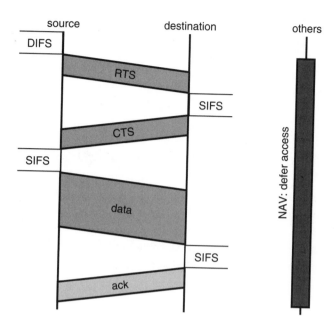

must use the standard transmission method to obtain access to send the RTS packet. Once the receiving host receives the packet, the host replies with a CTS message that includes the same duration of the session about to happen.

When node B replies with this CTS message, node C hears this response, is made aware of the potential collision, and will hold its data for the appropriate amount of time, preventing a collision. If every node on the network is using RTS/CTS, collisions are guaranteed to occur only while in the contention window.

RTS/CTS adds significant overhead to the wireless 802.11 protocol, especially at small packet sizes. If used, RTS/CTS thresholds must be set on both the access point and the client side.

Roaming

In a typical environment, two or more access points will provide signals to a single client. The client is responsible for choosing the most appropriate access point based on the signal strength, network utilization, and other factors. When a station determines that the existing signal is poor, it begins scanning for another access point. This can be done by passively listening or by actively probing each channel and waiting for a response.

Once information has been received, the station selects the most appropriate signal and sends an association request to the new access point. If the new access point sends an association response, the client has successfully roamed to a new access point (make-then-break behavior).

MAC Layer and Data Payload

In addition to collision avoidance, timing, and roaming, the MAC layer is also responsible for identifying the source and destination address of the packet being sent, as well as the data payload and a *cyclic redundancy check* (CRC). The entire payload of the packet, including the MAC header, is transmitted at the rate specified in the PLCP.

Home Networking

Figure 8-14 shows a cable modem connecting a home-based PC to the Internet using a 4-port router and a wireless access point. In this figure, one PC is connected to the router with a 10BaseT connection and a second device (laptop) is connected to the router through the wireless access point by radio frequency.

802.11b versus HomeRF Networks

A wireless system called 802.11b or Wi-FI is becoming very popular. It runs up to 11 Mbps, about as fast as a normal 10BaseT LAN and much faster than a 1.5-Mbps (maximum) cable modem. 802.11b sounds like a very techie term, but it isn't. It's just the name of a particular specification for wireless networking from a group that likes to use numbers for names, the IEEE Standards Association. They have many specifications with numbers for names. 802 is the *local area and metropolitan area networks standards* (LAN/WAN). Within that, 802.3 is Ethernet, 802.5 is Token Ring, and so on, and 802.11 is wireless LAN. 802.11b is the version referring to 2.4-GHz radio using DSSS and CSMA/CA that operates at 11 Mbps. The Wireless

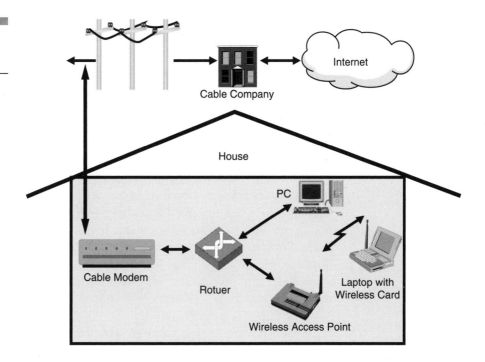

Figure 8-14
A wireless home network

Ethernet Compatibility Alliance is trying to get the name Wi-Fi accepted as a common term for 802.11b.

Conversely, HomeRF provides simultaneous support for up to eight toll-quality voice connections, eight prioritized streaming media sessions, and multiple Internet and network resource connections at broadband speeds. HomeRF accomplishes this with good comparative ratings for low-cost, small-size, low-power consumption, interference immunity, security, and support for high network density.

HomeRF has a 10-Mbps peak data rate in current products with sufficient range for most residential applications. The data rates slow to 5 Mbps or lower if needed to extend the range well outside the typical home. HomeRF also supports low-cost roaming for larger installations. HomeRF is the only technology with true toll-quality voice even in the presence of severe interference and the handset range is comparable to the best 2.4-GHz products on the market. It is currently the only technology with support for all CLASS service features like call waiting, caller ID, forwarding to individual handsets, distinctive ringing, and 911 breakthrough on an interoperable basis due to its DECT-based call stack. Unlike its 2.4-GHz band competitors, HomeRF's 2002 roadmap to 20 Mbps or higher with full backwards compatibility is not dependent on further FCC rule changes. Compared to HomeRF, Bluetooth lacks the capability to simultaneously support multi-line telephony, broadband speed data access, and multiple streaming sessions. 802.11 variants fail to provide toll-quality voice services and compare poorly in critical aspects such as cost, size, power consumption, and support for high network density. The serious shortcomings of 802.11 for interference immunity and security are a concern for mass-consumer deployment.

Whereas existing wire-based networks offer no-new-wires installation, wireless takes this one step further to enable no-wires convenience. A wireless home network uses RF instead of wires or cables to transmit information. A network unrestricted by wires is the most suitable solution for the home. Most homeowners do not want to incur the expense and inconvenience of running special cables throughout their homes, nor do they have a dedicated IT staff to assist with network installation. Wireless home networking avoids such problems, providing users with a low-cost, easy-to-use network that is simple to install.

Of the home networks available, wireless networks offer the most flexibility. They enable users to share peripherals and Internet access while also giving them mobility. For example, a wireless home network could enable a laptop user to download a file from the World Wide Web and print it on a printer connected to a networked PC in the family room—all from the comfort of his or her back patio. The most useful wireless home networks support the household's needs for voice as well as data. These networks enable users to talk, surf, compute, and communicate from anywhere in or around the home without being tied to telephone jacks or power outlets. Solutions are being developed that will enable multiple network environments to work together easily. For example, wireless networks will be designed to communicate seamlessly with phone line or Ethernet networks. This will soon enable home networks to serve as a backbone for interconnecting PCs, information appliances, home security systems, and digital entertainment centers.

802.11 has much in common with 802.11b, but it is legally scalable to much higher data rates (up to 54 Mbps) at the expense of significantly reduced range at its peak data rate. By operating at 5 GHz, 802.11a avoids much of the current interference concerns associated with 802.11b and mitigates the network density limitations of 802.11b. 802.11a will be more expensive than HomeRF or Bluetooth for many years to come and it fails to provide toll-quality voice services. It also suffers from the same security deficiencies as 802.11b.

Even though Bluetooth and some 802.11 variants will do well in their respective applications outside the home, this is not a compelling argument for service providers to compromise their revenue opportunities in the networked home. HomeRF is a solid choice for wireless networking of the broadband Internet home and benefits service providers and their customers. The world will adapt to the reality of multiple useful wireless connectivity interfaces for different applications, just as it already has adapted to a plethora of existing wired interfaces for the same reason.

Wireless Access Point

Wireless access points make part of the LAN wireless. The wireless access point connects to the LAN like any other device and then lets any computer connected to it wirelessly act as if it were on the same LAN. Combined wireless access points and routers are available to extend the reach of the LAN and limit the amount of boxes in use. Figure 8-15 illustrates a wireless access point.

Figure 8-15
A wireless access point
(Source: Linksys)

Units are configured by connecting a computer to them either with a USB or a serial connector. You need to set the name of your ESSID (your logical wireless LAN). The access points provide encryption using the *Wire Equivalent Privacy* (WEP) in their setup.

Wireless Adapter

To connect wirelessly, the PC or laptop needs a wireless adapter. The wireless adapters use a normal PCMCIA Type-II card that goes into the slot of a laptop just like a LAN card. The antenna sticks out. Adapters are available for desktop PCs that let you use the wireless PC cards. You plug the adapter boards into a PCI slot and then plug the same type of card you use on a laptop into that board. Figure 8-16 shows the PCMCIA card.

Realities of Wireless

Line-of-sight outdoors can work from quite a distance, but indoors, with walls and electrical devices, you get more interference and much less obvi-

Figure 8-16
The wireless interface for a laptop

ous coverage patterns. Other 2.4-GHz devices such as cordless phones and Bluetooth devices can cause temporary gaps in operation. When mounting the access point, higher is better. The higher the device is mounted, typically the fewer obstacles in the transmission path.

Some Motivation

The major motivation and benefit from wireless LANs is increased mobility. Untethered from conventional network connections, network users can move about almost without restriction and access LANs from nearly anywhere. Examples of the practical uses for wireless network access are limited only by the imagination of the application designer. In addition to increased mobility, wireless LANs offer increased flexibility. Again, imagination is the limiting parameter. One can visualize without too much difficulty a meeting in which employees use small computers and wireless links to share and discuss future design plans and products. This ad hoc network (shown in Figure 8-17) can be brought up and torn down in a very short time as needed, either around the conference table and/or around the world.

Figure 8-17
The ad hoc
wireless LAN

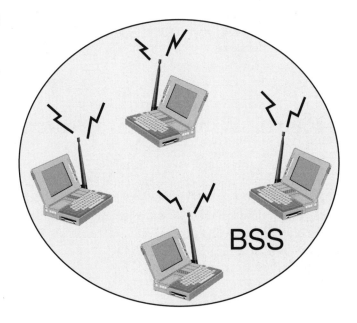

Mobile IP

Mobile *Internet Protocol* (IP) was suggested as a means to attain wireless networking. It focuses its attention at the network layer, working with the current version of the IP (IPv4). In this protocol, the IP address of the mobile machine does not change when it moves from a home network to a foreign network. In order to maintain connections between the mobile node and the rest of the network, a forwarding routine is implemented.

When a person in the real world moves, he or she lets the home post office know to which remote post office his or her mail should be forwarded. Upon arriving at his or her new residence, the person registers with the new post office. This same operation happens in Mobile IP. When the mobile agent moves from its home network to a foreign (visited) network, the mobile agent tells a home agent on the home network to which foreign agent its packets should be forwarded. In addition, the mobile agent registers itself with that foreign agent on the foreign network. Thus, the home agent forwards all packets intended for the mobile agent to the foreign agent, which sends them to the mobile agent on the foreign network. When the mobile agent returns to its original network, it informs both agents (home and foreign) that the original configuration has been restored. No one on the outside networks needs to know that the mobile agent moved.

This configuration works, but it has some drawbacks. Depending on how far the mobile agent moves, some packets may need to be stored and forwarded while the mobile agent is not on the home or foreign network. In addition, Mobile IP works only for IPv4 and does not take advantage of the features of the newer IPv6.

Faster Wireless Standards: 802.11a

The 802.11a standard, on the other hand, was designed to operate in the more recently allocated 5-GHz *Unlicensed National Information Infrastructure* (UNII) band. Unlike 802.11b, the 802.11a standard departs from the traditional spread spectrum technology, and instead uses a frequency-division multiplexing scheme that's intended to be friendlier to office environments.

The 802.11a standard, which supports data rates of up to 54 Mbps, is the Fast Ethernet analog to 802.11b, which supports data rates of up to 11 Mbps. Like Ethernet and Fast Ethernet, 802.11b and 802.11a use an identical media access control (MAC). However, whereas Fast Ethernet uses the same physical-layer encoding scheme as Ethernet (only faster), 802.11a uses an entirely different encoding scheme, called *orthogonal frequency division multiplexing* (OFDM).

Frequencies for All

The 802.11a standard is designed to operate in the 5-GHz frequency range. The FCC allocated 300 MHz of spectrum for unlicensed operation in the 5-GHz block. Table 8-2 shows the spectrum allocated by the FCC.

The total bandwidth allotted for IEEE 802.11a applications is almost four times that of the ISM band; the ISM band offers only 83 MHz of spectrum in the 2.4-GHz range, whereas the newly allocated UNII band offers 300 MHz. The 802.11b spectrum is plagued by saturation from wireless phones, microwave ovens, and other emerging wireless technologies, such as Bluetooth. 802.11a spectrum is currently relatively free of interference due to the lack of devices operating in that spectrum. This may change over time, but the overall opinion is that it is better than to be in the congested areas of 2.4 GHz.

The 802.11a standard gains some of its performance from the higher frequencies at which it operates. Moving up to the 5-GHz spectrum from 2.4 GHz, however, will lead to shorter distances. In addition, the encoding mechanism used to convert data into analog radio waves can encode one or more bits per radio cycle (hertz). Power is increased to compensate for loss in the frequencies over the distances. Power by itself is not enough to maintain the same distances in an 802.11b environment. A new physical-layer encoding technology that departs from the traditional direct-sequence technology is being used. This technology is called *coded OFDM* (COFDM). COFDM was developed specifically for indoor wireless use and offers performance much superior to that of spread spectrum solutions. COFDM works by breaking one high-speed data carrier into several lower-speed subcarriers, which are then transmitted in parallel. Each high-speed

Table 8-2

Spectrum Allocated
for 802.11a by
the FCC

Allotted Bandwidth	Operating Frequency Band
200 MHz	5.15–5.35 MHz
100 MHz	5.725–5.825 MHz

■■■ ■■■ ■■■ ■
Figure 8-18
The 52 subchannels
per carrier

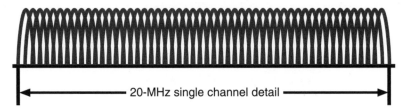

Subchannels

52 carriers per channel

◄──────── 20-MHz single channel detail ────────►

Each channel is subdivided into 52 sub channels, each about 300 KHz wide.

■■■ ■■■ ■■■ ■
Figure 8-19
Independent clear
channels

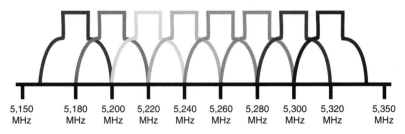

Independent Clear Channels

| 5,150 MHz | 5,180 MHz | 5,200 MHz | 5,220 MHz | 5,240 MHz | 5,260 MHz | 5,280 MHz | 5,300 MHz | 5,320 MHz | 5,350 MHz |

*There are eight independent clear channels in the lower 200 MHz
of the 5-GHz spectrum.*

carrier is 20-MHz wide (Figure 8-18) and is broken up into 52 subchannels, each approximately 300-KHz wide (Figure 8-19). COFDM uses 48 of these subchannels for data, while the remaining four are used for error correction. COFDM delivers higher data rates and a high degree of multipath reflection recovery, thanks to its encoding scheme and error correction.

One of two pioneers developing 802.11a says it will support data rates of 6 Mbps, 12 Mbps, and 24 Mbps, as per the standard. It will also support data rates of 36 Mbps, 48 Mbps, and 54 Mbps. Others will follow and support the same variety of data rates.

The de facto standard for 802.11a networking appears to be 54 Mbps. Data rates of 54 Mbps are achieved by using 64QAM. Vendors are now stating that they will offer an additional proprietary mode that combines two carriers for a maximum theoretical data rate of 108 Mbps. They conservatively estimate that data rates of 72 Mbps will be possible when using its proprietary dual-channel mode.

Bran Anyone?

In Europe, the HiperLAN/2 standard led by the *European Telecommunications Standards Institute* (ETSI) called *Broadband Radio Access Networks* (BRAN) group has wide acceptance as the 5-GHz technology of choice. HiperLAN/2 and 802.11a share some similarities at the physical layer. However, HiperLAN/2 is more like *Asynchronous Transfer Mode* (ATM) than Ethernet. The HiperLAN/2 standard actually grew out of the effort to develop wireless ATM. HiperLAN/2 shares the 20-MHz channels in the 5-GHz spectrum in time, using *time-division multiple access* (TDMA) to provide *Quality of Service* (QoS) through ATM-like mechanisms.

802.11a, on the other hand, shares the 20-MHz channel in time using CSMA/CA. Logically, HiperLAN/2 uses a different MAC than 802.11a. The HiperLAN/2 MAC design has proven to be problematic and controversial, and the HiperLAN/2 standard is not yet complete. In contrast, 802.11a uses the same MAC as 802.11b, which gives developers only one task to complete: a 5-GHz IEEE 802.11a-compliant radio. This is no simple task, but it is easier than redesigning the radio and the MAC controller.

Taming the Standards Beast

Manufacturers are concerned over the competing 802.11a and HiperLAN/2 standards. Building and supporting two separate products is a nightmare in both development and marketing. The increased development costs will be handed down to the end user, which will detract from the acceptance of the standards and products. This is a lose-lose situation for the vendors. A vendor proposed standard, called 5-*Unified Protocol* (UP), provides extensions to 802.11a and HiperLAN/2, letting both technologies interoperate at low, medium, and high speeds. The 5-UP standard also specifies a method for selecting subchannels for transmission within a carrier. If this portion of 5-UP were adopted, it could enable devices such as wireless phones, Bluetooth products, and other narrow-bandwidth applications to use a part of the 5-GHz spectrum without having a significant impact on network performance. This would help prevent the saturation and congestion problems that have arisen in the 2.4-GHz space. The judge and jury are still out on the decision as of this writing; the IEEE has the proposal under advisement.

Interoperability Problems

Because 802.11a and 802.11b operate in different frequencies, there's no chance they'll be interoperable. Thus, any vendor or end user that has made large investments in 802.11b technology faces the forklift technology dilemma. The only way to change the architecture is to bring in the forklifts and take away the 802.11b, then bring in the 802.11a boxes. A migration path exists for the future when more bandwidth is required, but extensive retooling is required. The 802.11a and 802.11b technologies can coexist, however, because no signal overlap occurs. Let's wait and see where this one leads.

Dental Hygiene Anyone?

When we look at the wireless standards, the idea of creating an open standard that will address all of our wired and wireless needs always crops up. Bluetooth is an example of these types of standards that are built upon the idea of broadband wireless communications and wired communications.

What Is Bluetooth?

Bluetooth is a global standard that

- Does away with the wires and cables between stationary and mobile devices
- Handles data and voice communication services
- Allows the creation of ad hoc networks
- Synchronizes all of our devices

The Bluetooth wireless technology consists of hardware, software, and the capability to interoperate among devices. Just about all-major players in the telecommunications, computer, and home-entertainment industry have adopted the standards. Moreover, the Bluetooth standard is accepted throughout various sectors of the industry like automotive, health care, entertainment, and office automation. As the industry has seen the explosion in the information services (IS) sector, mobility in people's business and personal lives has become paramount. More people use mobile communications at a faster growth rate than wired communications services. Bluetooth is the enabling wireless technology that facilitates interpersonal communications among people by means of a low-cost, short-range radio link.

Bluetooth Roots

The idea that resulted in the Bluetooth wireless technology was born in 1994 when Ericsson Mobile Communications decided to investigate the feasibility of a low-power, low-cost radio interface between mobile phones and their accessories. The intent was to use a small radio built into both the cellular telephone and added to a laptop replacing the need for a cable to connect the two devices. Within a year, engineering work began to reveal the true potential of the technology. However, besides untethering these devices and eliminating the cables, the radio technology offered the opportunity to universally bridge across existing data networks, provide a peripheral interface, and facilitate the formation of the private ad hoc groups.

In February 1998, the *Special Interest Group* (SIG) was formed to build upon the standards and support the development of products and services. Bluetooth is a specification for a radio solution, which is small form-factor, low cost, and low power, that provides seamless wireless connectivity between notebook computers, cellular phones, and other portable handheld devices. The Bluetooth SIG is comprised of leaders in the telecommunications, computing, and networking industries that are driving the development of the technology to bring it to market. The Bluetooth SIG includes promoter companies such as 3Com, Ericsson, IBM, Intel, Lucent, Microsoft, Motorola, Nokia, Toshiba, and over 2,000 other adopter/associate member companies. Bluetooth operates in the license-free 2.4-GHz ISM band at a link range of 10 meters. With improved transmission power and receiving sensitivity, the range can be increased up to 100 meters. Bluetooth is expected to be a standard feature on the next generation Smart Phones, PDAs, and computers. The most compelling application for Bluetooth is the always-on Internet access at homes, offices, and public locations through a Bluetooth Internet Access Point. Users will access the Internet with their Bluetooth-enabled notebook, PDA, or Smart Phone.

From the onset, the main goals for the SIG have been to include a regulatory framework in the specification that guarantees full interoperability between different devices from various manufacturers, as long as they share the same profile. The specification lists several different protocol stacks and profiles. Different applications may run on top of different protocol stacks. Moreover, a number of usage models are identified and will show how the protocols are stacked to support the usage model.

Whereas the usage models describe applications designed for specific devices, the profiles specify how to use the Bluetooth protocol stack for an interoperable solution. Each profile outlines how to reduce options and set parameters in the base standard while setting the procedures and usage from several base standards. A common user experience is also defined. A mouse or keypad doesn't need to communicate with a headset; therefore, they are built to comply with different profiles. The profiles are a part of the Bluetooth Specification. All devices are tested against one or more of the profiles to qualify for Bluetooth certification. Figure 8-20 shows the complete Bluetooth protocol stack.

Compliance

The Bluetooth Qualification Program guarantees global interoperability between devices regardless of the vendor or the country of use. The test procedures, which all devices must pass, verify that the devices meet all

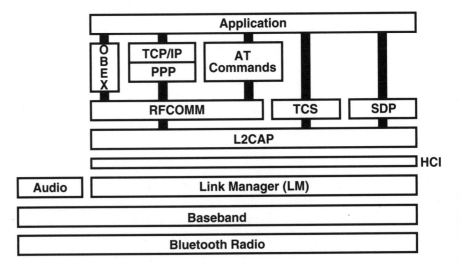

Figure 8-20
Bluetooth protocol stack

requirements regarding radio-link quality, lower-layer protocols, profiles, and information to end users.

Countless electronic devices for home, personal, and business use have been presented to the market during recent years, but no widespread technology is available to address the needs of connecting personal devices in *personal area networks* (PANs). The demand for a system that could easily connect devices for the transfer of data and voice over short distances without cables grew stronger.

Bluetooth wireless technology fills this important communication need, with its capability to communicate both voice and data wirelessly, using a standard low-power, low-cost technology that can be integrated in all devices to enable total mobility. The price will be low and result in mass production. The more units that are around, the more benefits for the customer.

The profiles defined in version 1 of the Bluetooth Specification mainly address usage models concerning the telecom and computing industries. The Bluetooth Specification defines a short (around 10 m) or optionally a medium range (around 100 m) radio link capable of voice or data transmission up to a maximum capacity of 720 Kbps per channel.

Radio frequency operation is in the unlicensed *industrial, scientific, and medical* (ISM) band at 2.4 to 2.48 GHz, using a spread spectrum, frequency-hopping, full-duplex signal at up to 1,600 hops/sec. The signal hops among 79 frequencies at 1 MHz intervals to give a high degree of interference immunity.

Voice

Up to three simultaneous synchronous voice channels, or a channel that simultaneously supports asynchronous data and synchronous voice, are used. Each voice channel supports a 64-Kbps synchronous (voice) channel in each direction.

Data

The asynchronous data channel can support maximal 723.2 Kbps asymmetric (and up to 57.6 Kbps in the return direction) or 433.9 Kbps symmetric.

■ A master can share an asynchronous channel with up to seven simultaneously active slaves in a piconet.

- By swapping active and parked slaves out respectively in the piconet, 255 slaves can be virtually connected.
- There is no limitation to the number of slaves that can be parked.

Bluetooth units that come within range of each other can set up ad hoc point-to-point and/or point-to-multipoint connections. Units can dynamically be added or disconnected to the network. Two or more Bluetooth units that share a channel form a piconet.

Several piconets can be established and linked together in ad hoc scatternets to enable communication and data exchange in flexible configurations. If several other piconets are within range, they each work independently and have access to full bandwidth. A different frequency-hopping channel establishes each piconet. All users participating on the same piconet are synchronized to this channel. Unlike infrared devices, Bluetooth units are not limited to line-of-sight communication.

Two other short-range radio technologies using frequency-hopping techniques reside in the 2.4-GHz band:

- Wireless LANs based on the IEEE 802.11 standard. The technology is used to replace a wired LAN throughout a building. Both the transmission capacity and the number of simultaneous users are high. On the other hand, compared to Bluetooth wireless technology, it is more expensive and power consuming, and the hardware requires more space. It is therefore not suitable for small mobile devices.
- The other 2.4-GHz radio is HomeRF, which has many similarities to the Bluetooth wireless technology. HomeRF can operate ad hoc networks (data only) or be under the control of a connection point coordinating the system and providing a gateway to the telephone network (data & voice). The hop frequency is 8 Hz, whereas a Bluetooth link hops at 1,600 Hz.

The 802.11 family of products will be a strong competitor for the PC and notebook market segment, but others feel that Bluetooth will likely be the standard wireless technology supported by next generation PDAs and Smart Phones due to its lower cost, lower power consumption, and industry commitment. Table 8-3 shows the competing technologies.

Sales of Smart Phones and PDAs are growing at an ever-increasing rate, and Bluetooth is the wireless standards-based solution for these platforms. We believe that Bluetooth-enabled clients will be pervasive due to the compelling personal area networking applications. With this ubiquity of Bluetooth and with a global trend towards Mobile Internet Access, Pico

Table 8-3

Bluetooth
Competing
Technology

Technology	Data Rate Mbps	Range Meter	Frequency GHz	Specifi-cation	Status
802.11	2	100	2.4	1999	Now
802.11b	11	100	2.4	1999	Now
802.11a	54	TBD	5	2000	Soon
802.15 (Bluetooth)	1	10	2.4	1999	Now
802.15(High rate)	20+	TBD	2.4 or possibly 5	TBD	N/A
HomeRF	1.6	50	2.4	1999	Now
HomeRF (Next Gen)	10	50	2.4	N/A	FCC Approve
IrDA	4	1	Infrared 100 THz	Now	Now

communications will be solidly positioned to provide the equipment and infrastructure management tools to enable Mobile Internet Access.

Ultra-Wideband Radio (UWB) is a new radio technology still under development. Short pulses are transmitted in a broad frequency range. The capacity appears to be high, whereas power consumption is expected to be low.

Wireless LANs will become equally important to our roaming needs and our mobility needs for the future. One can see how the mobile and the wireless communities are demanding speeds that are faster and more reliable. We need to get the throughout to the next level of reliability and speed to gain the acceptance. Once this level of acceptance is achieved, we will see the convergence of the LAN and WAN, especially the Internet integration for the future.

The judge and jury are still in the wings waiting to see the capabilities of the 54 Mbps speeds on a wireless LAN, then the fixed wireless access to the outside to the Internet for the next step over the next 2 to 3 years.

Wireless Innovations in Broadband

Wireless Innovations

Most of what we have discussed to date has been regarding the technologies used to handle the wireless communications. The LAN architectures led us close to other issues such as the protocols used to support the high-speed and broadband communications needs for the future. It is a natural propensity to discuss the hardware components used to produce products and services. However, many other innovations are taking place all around the hardware. These innovations include the use of communications protocols to handle the preparation and delivery of the data communications. Note that the issue hangs on data delivery. The use of data is a generic statement because we use digital communications that turns voice and data into strings of ones and zeros for transmission. This means that voice is data and data is too! However, we must also look at the other side of this coin. Voice is non-granular—if we add broadband communications, we do not get better quality for voice because the 64-Kbps *pulse coded modulation* (PCM) voice stream is all we will ever need. Indeed, 64 Kbps is overkill to handle a single voice call. Thus, we concentrate on broadband communications to move large amounts of data, multimedia, and video capabilities rather than voice. Having said that, the next step is to look at the innovations of our data protocols to support the high-speed communications needs for the future.

The Market in General

Wander through any airport, office building, or shopping mall and you will pass dozens of fellow business people with cell phones glued to their ears. Look into any gate area or airport lounge, and you can't miss the road warriors with laptops propped open on their briefcase or on bent knee. Many of these road warriors have a data transfer ongoing using some form of wireless connection. Many will have an internal modem; others will be using an external cell phone plugged into their laptop modem card. More people yet will have a *personal digital assistant* (PDA) (like a Palm) connected to some form of wireless network. Finally, those WAP-enabled cell phones will be logged onto e-mail or browsing a special Web site. After takeoff, the buzz of the familiar Windows™ jingle occurs as these businesspeople resume computing for the rest of the flight (that excludes the people who play games).

Increasingly, we live in a world requiring a non-stop communication and computing environment. Laptops and subnotebooks are the norm in corporate America. These were once considered the status symbol of the geek. Now the wireless modem and full access to e-mail on-the-fly are the services to die for.

No wonder entrepreneurs around the world are welcoming the convergence of cellular communications and computer technologies creating *wireless network computing*. How often we have seen people linking (networking) their PCs together in a plane with the Infrared port (IrDa) on their laptop so they can synchronize files or calendars. This is such a common occurrence today it is becoming somewhat invisible. However, using a Lap Link™ application to synchronize files between two coworkers is minimally a wireless data networking strategy. Linking similar files and applications together over a much wider geographical distance requires more complex interaction and protocols.

From Suitcase to Palmtops

Portable computers first appeared as what was called a laptop. The only problem is that no one could determine on whose lap this behemoth would comfortably sit. These devices were so big and heavy it took a suitcase to cart them around. They have since shrunk considerably to the current size of a notebook (or less) and weigh a mere 1 to 2 lbs. Subnotebooks and even *palmtops* became the envy of many people. The proliferation of palmtop devices (from 3Com, Vigor, HP, Compaq, and others) has steadily created the connected image from the handheld device. With these devices ranging from a couple hundred dollars on up, the novelty of communicating back to the office became contagious. No longer do we want to communicate back to the home office while traveling, we must do so.

Speed kills, except when we are talking about the capacity of the notebooks and subnotebook computers. Processor speed, storage, and memory have increased, driving the market to a feeding frenzy. In the late 1990s, mobile communications became necessary because many of the vendors were shipping *Cellular Digital Packet Data* (CDPD), wireless LAN cards, or wireless modems along with the notebooks and palm devices. Today, the devices that cost over $300 to $500 to equip a notebook have fallen to less than $100. The market is slowly flooding with service offerings from the top suppliers as well as many startup companies. Availability has become the driver for the demand. With so many choices, it is only natural that everyone wants one. However, the growing number of road warriors using these handheld devices is placing a burden on the wireless infrastructure and the protocols that support the wireless transmission.

On the flip side of the coin, cellular phones are smaller, lighter, cheaper, and smarter. WAP enabled, they are now using our wireless data networks to support applications such as Internet e-mail, corporate e-mail, chat, and text-based browsers. *Global System for Mobile Communications* (GSM) cellular phones and personal organizer combinations can send e-mail and access Web pages in *Tagged Text Markup Language* (TTML) format. AT&T's wireless cell phone and its PocketNet is an IP-enabled CDPD phone. The phone is equipped with a small LCD screen that can be used to access e-mail, online services, and Web pages in *Handheld Device Markup Language* (HDML) format. The service is slow (9.6 to 19.2 Kbps), but it is relatively inexpensive (free access in country).

Motorola's Accompli shown in Figure 9-1 integrates e-mail, *short message service* (SMS), and other functions from a two-way communicator at

Figure 9-1
Motorola Accompli
(Source: Motorola)

slow speeds today, but will soon be *general packet radio systems* (GPRS) compatible and support 28.8 to 170 Kbps. This is not yet a broadband service, but it does build the architecture to take the next step in speed. The *Enhanced Data for GPRS Environment* (EDGE) will support 384 Kbps. However, many carriers may choose to skip by this step and go straight to the *Universal Mobile Telephone Service* (UMTS) and 2-Mbps speeds.

All these devices provide a hardware and operating system foundation upon which wireless data networks and applications build. This handheld battalion of products continues to shrink in size, yet increases are escalating in memory and modem speed. Improvements make these handheld devices more attractive and affordable for the masses. Converging voice, data, fax, multimedia, and paging services is essential to continue the proliferation of speed and service. The days of carrying different devices for different services are becoming passé. Work also continues on the variation of the H.324 standard for video conferencing systems to support wireless videophones (the Dick Tracy telephones).

Metricom is an emerging wireless provider attempting to provide a faster, low-cost alternative to nationwide wireless network services. Metricom uses a combination of pole top radios and wired access points, strategically located throughout a relatively small area such as a campus or urban business center. Metricom's technology is proprietary and requires use of Metricom's Ricochet modem. A gateway must be used to obtain Internet access through Metricom's network. Recent enhancements support up to 128 Kbps for packet data transmission.

A wireless modem is slightly more expensive than an analog PCMCIA modem. However, the true cost issue for most subscribers will be usage sensitive charges. It may be difficult to justify the expense of transmitting wireless data when it is usage sensitive, especially with the downturn in economic conditions. One price for "all you can eat" plans will be effective leverage for the providers.

Wireless data network services are proprietary and do not provide sufficient interoperability. Exceptions to this quandary are CDPD and the RAM Mobile Data, both based on standard network protocols. A proprietary wireless infrastructure may be cheaper or faster. If you buy a CDPD modem, you can change service providers. The same cannot be said of buying a Ricochet modem.

Ricochet delivers speed, security, reliability, and ubiquity. It was one of the first mobile solutions to duplicate high-speed wired access, enabling people to work outside the confines of their office. It provides mobile data access that is efficient, reliable, and compatible with existing hardware, software, and industry protocols. Built from the ground up to handle mobile access, Ricochet was developed by Metricom. Leveraging patented technology, radio frequencies, and IP-based networking; Ricochet offers high-speed mobile data access and interoperability with current network technologies. This means mobile professionals using notebook computers can access online information exactly as if they were at their desktop. They can view and download full Web content, send and receive e-mail with attachments, link to corporate intranets through *Virtual Private Networks* (VPNs), and use the applications, information, and setup they're accustomed to without needing to be in their office. It's reliable mobile access without the wires. It's true mobile freedom for the first time.

The Ricochet Network Architecture

The network is a combination of several individual elements. This includes subscriber devices, microcell radios, wired access points, network interconnect facilities, gateways, name server, and a network operations center. Figure 9-2 shows these components.

The Components

The pieces of the Ricochet network are described herein so that we can see the differences of using the proprietary networking architecture as opposed to using an open standard like CDPD. The components are required in order to operate on the Ricochet network but incompatible with other networking operations.

- **Ricochet wireless modems** Ricochet modems are compact, portable, rugged, and easy to install. All work with a variety of PC, Macintosh, and handheld devices. They use USB, standard serial, and PC Card (PCMCIA) interfaces.

  Acquisition/authorization: Acquisition is a necessary first step for each radio on the network. When a radio is turned on, it locates neighboring radios and Ricochet modems by sending out synchronization packets. Once it receives acknowledgement packets from neighboring radios and authorization from the name server, it becomes a fully functional member of the network.

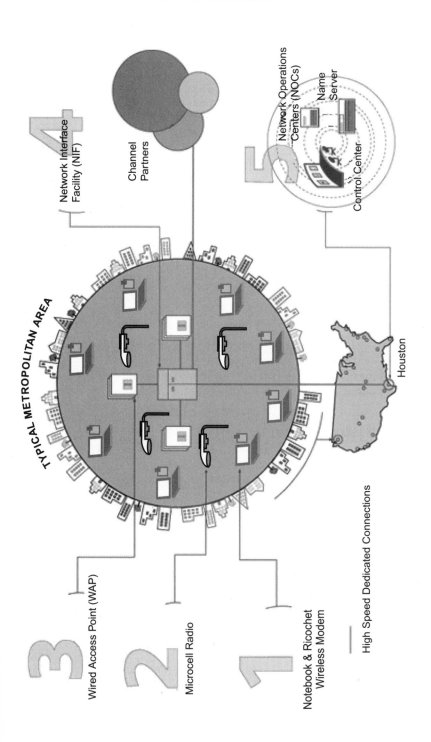

4 Network Interface Facility (NIF)

Channel Partners

5 Network Operations Centers (NOCs)

Name Server

Control Center

TYPICAL METROPOLITAN AREA

Houston

3 Wired Access Point (WAP)

2 Microcell Radio

1 Notebook & Ricochet Wireless Modem

— High Speed Dedicated Connections

Figure 9-2 Components of Ricochet network (Source: Metricom)

■ **Microcell radios** These intelligent routing, self-contained, shoebox-sized radio transceivers are typically mounted to streetlights or utility poles approximately five per square mile in a mesh network. Microcell radios communicate with users' laptops or handheld devices through Ricochet wireless modems. Microcells are configured to route incoming packets to the optimal wired access point.

Acknowledgement and error checking: Every packet sent contains a compressed summary of the data that can be used to confirm the accuracy of the received packet. To ensure reliable transmission of data, the receiving radio verifies that the received data is correct and then it sends a brief *acknowledgement* (ACK) data packet back to the originating radio to verify that the packet was sent correctly.

■ **Wired Access Points (WAPs)** Strategically placed within a 10- to 20-square mile area, each WAP collects and converts the radio frequency packets into a format for transmission on a local wired IP network.

An integral part of the WAP is the Ethernet radio that operates just like a microcell radio but includes Ethernet hardware and software capabilities. Ethernet radios convert radio data packets into a format for transmission to a wired IP network. A WAP also contains standard IP networking equipment.

The WAPs convert radio data packets to protocols that can be used on a nationwide, wired network. This exchange takes advantage of the faster speeds inherent in wired-line technologies, such as fiber options and T-1 connections that are all part of the Metricom high-speed wired network.

■ **Network Interface Facility (NIF)** All WAPs in a region are connected to the regional NIF. The NIFs house gateways that connect the Ricochet wired network to the *Network Operations Center* (NOC) and other major networks of Metricom's channel partners.

■ **Ricochet Gateway** A Ricochet Gateway consists of hardware and software designed to connect the Ricochet network with other networks. Its function is to remove the Ricochet protocol and route the end user's IP packets to their specified destination.

■ **Network Operations Center (NOC)** Two highly scalable NOCs serve the entire Ricochet system. Operating simultaneously, the two NOCs split the load across the country. If one of the NOCs needs an upgrade or experiences an outage, the other NOC will take over all traffic. The NOCs are located in Dallas and Houston.

A wired network links the Operations Centers with all the NIFs and through them to the WAPs, radios, and subscriber devices of the wireless network. Every microcell radio, Ethernet radio, IP router, and communication circuit is monitored and maintained from the NOC using Metricom's *Network Management System* (NMS).

■ **Name Server**   The Ricochet Name Server maintains access control and routing information for every radio and service within the Ricochet network. Every time a Ricochet device (subscriber device, microcell radio, or gateway) is powered on, it registers with the Name Server to verify that it has network authorization. Whenever a Ricochet device requests a connection, the Name Server validates the request. If authorized, the originator is provided with a network routing path to the requested destination. This feature adds yet another level of security for all individuals and organizations using the Ricochet network.

Radio Frequency Spectrum

To reduce latency and increase performance, Ricochet incorporates multiple *radio frequency* (RF) bands for full-duplex transmission between WAPs, pole tops (intelligent microcell radios), and the high-speed end-user modems. Two *Industrial Scientific Medical* (ISM) bands of regulated, unlicensed spectrum are used: the 900-MHz band and the 2.4-GHz band in addition to the licensed 2.3-GHz *Wireless Communications Systems* (WCS) spectrum.

Frequency Hopping, Spread Spectrum Technology

Each microcell radio uses many frequency-hopping channels in the 900-MHz and 2.4-GHz bands and uses a randomly selected hopping sequence, enabling multiple subscribers to use the network simultaneously. This technique makes more efficient use of a given band of frequencies because many more radio transmissions can take place simultaneously without interfering with one another. It is also extremely robust and will easily cut through potential interference.

Packet-Switched Networking

For efficient transfer of data over radio waves, the Ricochet network transmits data by segmenting and routing information in discrete data units called *packets*. Each packet has its own control information for routing, sequencing, and error checking. Packets are routed from radio to radio across the wireless network. This means that multiple communications between radios can occur simultaneously. With packet switching, a communications channel can be shared by multiple users, with each using the channel only for the time required to transmit a single packet—typically no more than 20 milliseconds. In conjunction with frequency hopping, packet switching provides the Ricochet network with a tremendous amount of network capacity.

The basic structure of the Ricochet network consists of microcell radios strategically placed every quarter to half mile in a checkerboard pattern. This mesh architecture, as seen in Figure 9-3, routes data traffic between the modems and the WAPs. A higher density design like this ensures the safe, efficient transmission of data and better indoor penetration of the radio signals. It also offers the reliability and handoff capability necessary to eliminate dropped connections even when moving.

This mesh architecture provides advantages over the more typical network topology, known as the *star topology,* in which all communications are required to pass through one or more central base stations or hubs. In this system, congestion and impaired signal communications resulting from weak signal strength are generally addressed by installing another hub, typically a costly and time-consuming process. With the Ricochet network, we can reduce system congestion and increase network coverage and capacity by installing one or more relatively inexpensive pole top or wired access points where needed.

Middleware, Custom Protocols, and Proxies

Early attempts at wireless computing used *Transmission Control Protocol / Internet Protocol* (TCP/IP) directly over wireless data services to offer common applications such as e-mail. The slow speeds and lengthy delays associated with wireless data transfer prevent this solution from working very effectively. Yet, this approach may still be viable for faster, reliable

Figure 9-3
Mesh architecture
(Source: Metricom)

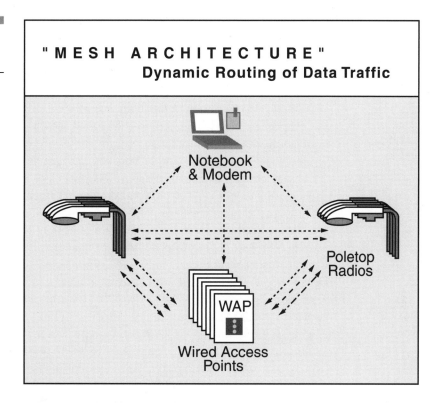

"MESH ARCHITECTURE"
Dynamic Routing of Data Traffic

Notebook & Modem

Poletop Radios

WAP

Wired Access Points

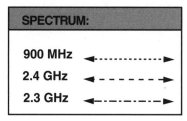

SPECTRUM:

900 MHz

2.4 GHz

2.3 GHz

wireless data services. Consequently, several alternatives emerged. Among these were

■ Custom data transport protocols, tuned for operation over a wireless data link or network

■ Customized implementations of *Transmission Control Protocol* (TCP), tweaked to improve operation over wireless IP

■ Middleware that insulates applications from wireless network characteristics

The custom alternative has been evaluated for several years without much success. Mobile IP addresses only part of the problem—enabling mobility of IP addresses. Myriad options and Internet proposals attempting to define custom transports have not yielded a standard. The enormous embedded base represented by TCP simply overwhelms the marketability of transport protocols engineered specifically for wireless.

Companies have implemented the second choice (custom tuning of TCP) with some success. End users cannot tune standard off-the-shelf TCP implementations to completely address the issues surrounding effective transmission over wireless. Custom TCP services require tuning at both ends for this to work. This requires some due diligence on the part of users.

The third choice, wireless network middleware, addresses this concern by constraining wireless adaptation to a (smaller) set of devices, known as *proxies*. Wireless clients interact with proxies using a middleware protocol, which provides reliable, efficient delivery over commercial off-the-shelf *User Datagram Protocol / Internet Protocol* (UDP/IP). The *Handheld Device Transport Protocol* (HDTP) used by AT&T's PocketNet is an example of this approach. Others exist, but one must know which protocol proxy is being used by the devices. No single solution is available to solve this problem.

The most common use of protocols today happens to be IP. We are all looking to use the broadband communications networks to enable our applications through the Internet and more specifically using IP. Usually we hear about transmitting data in the form of TCP and IP; other real-time protocols use UDP and IP. Thus we will look at TCP over wireless conventions and mobile IP applications using the protocol. Then we will tie these into the areas we have already covered such as Bluetooth, HomeRF, and WAP. This will give us a well-rounded approach to using broadband wireless transport for our broadband-hungry applications using underlying architecture to service these applications.

By casting this approach in concrete, we begin our venture looking at TCP and IP over wireless architecture. Currently, TCP assumes a relatively reliable underlying network where most packet losses are due to congestion. In a wireless network packet, losses will occur more often due to unreliable wireless links than due to congestion. When using TCP over wireless links, each packet loss on the wireless link results in congestion control measures being invoked at the source. Significant performance degradation occurs as a result. We must understand the effect of

- Bursts of errors on wireless links
- Packet size variation on the wired network
- Local error recovery taken by the base station
- Explicit feedback by the base station

The performance of TCP is sensitive to the packet size, and significant performance improvements are obtained if the best packet size is used. Local recovery by the base station using link-level retransmissions improves performance. However, timeouts can still occur at the source, causing redundant packet retransmissions. An *explicit feedback* mechanism to prevent these timeouts during local recovery solves this problem. When explicit feedback from the base station is used, the performance improvements are significant.

Mobile IP

Mobile IP provides a network layer solution to the mobility problem, specifically for the Internet. It is independent of the underlying protocols at the data link and physical layers. Therefore, mobile IP can be employed over a variety of media choices including the use of wireless communications. Mobile IP addresses the use of a portable address similar to a person who is going on an extended vacation. The person going on vacation notifies the *post office* (PO) that he/she will be gone for several months. The new address for this individual is passed on to the postmaster in the local PO. The PO will forward all new incoming mail to the new address. When our vacationer returns, the mail stops being forwarded and the local PO is responsible to deliver the mail.

A similar process occurs in mobile IP. The mobile node (wireless data terminal device) is provided with a temporary address at the new location (foreign network or serving system). The mobile can obtain a temporary address from a foreign agent (an *Interworking Function* [IWF] of serving system) that provides these services or from another source. The mobile registers its temporary address with its home agent (IWF of home system) on its home network. A home agent keeps track of the current location of the mobile node.

When data packets are sent to the mobile node, they arrive at the home network. The home agent intercepts the packets and reroutes them to the mobile node's current location. Thus, the mobile node IP solution enables the data terminal to communicate whereas other nodes are unaware of the current location of the node, even though they can communicate with the mobile node. A mobile node sends a packet directly without routing it through its home network. The basic characteristics of mobile IP include the following:

■ The mobile IP solution is provided at the network layer (IP). Therefore, it can be used on any data link layer (such as Ethernet or wireless).

- Mobile IP provides the capability to move without breaking connections. This happens because the node communicating with another node can still send data using the mobile's permanent IP address, and packets will be delivered to the mobile's current location.

- Mobile IP does not require significant changes to existing IP protocols.

- No changes are required to the infrastructure.

Mobile IP is, therefore, transparent to the users. With the exception of initial address acquisition, the user need not be aware of the mobile's location. The Mobile IP working group develops the standards for Mobile IP. Table 9-1 shows a list of these standards.

TCP/IP over Satellite

A fair amount of press has focused on TCP/IP, the networking protocol that the Internet and the *World Wide Web* (WWW) rely on, for not working properly over satellite transmissions. Satellite can support very high-speed communications such as multimegabit speeds. The claims are that the TCP/IP reference implementation's 4K buffer size limits channel capacity and data throughput to only 64 Kbps. The argument is then used that the maximum buffer of 64K limits maximum throughput to 1 Gbps. Therefore, the argument continues that GEO Ka-band satellite services are unsuitable for

Table 9-1

List of Mobile IP Standards

Standard	Description
RFC 2005	Applicability statement for IP mobility support
RFC 2002	Defines the Mobile IP protocol
RFC 2003	IP encapsulation with IP packets
RFC 2004	Minimal encapsulation within IP
RFC 1701	*Generic Routing Encapsulation* (GRE) methods
RFC 2006	*Management Information Base* (MIB) for mobile IP
RFC 2344	Reverse tunneling for mobile IP
RFC 2356	SKIP Firewall Traversal for mobile IP

high-bandwidth applications as the increased latency of a GEO connection decreases the available bandwidth.

TCP buffers are sized as

$$\text{bandwidth} * \text{delay} = \text{buffer size}$$

With a limited buffer size, a longer end-to-end delay decreases the space available to hold spare copies of unacknowledged data for retransmission. This limits the throughput on a lossless TCP connection.

However, this argument ignores the work done on larger buffer sizes for TCP in RFC 1323, the *large windows* effort. Work to expand TCP beyond its original 16-bit buffer space has been going on for several years. Moreover, the effort and result of large windows is supported by several versions of UNIX. The TCP buffer limit isn't as bad as it's made out to be. TCP copes with GEO delays quite nicely today. Individual high-bandwidth GEO TCP links are possible with the right equipment and software. Military applications have been using TCP/IP in many GEO-based networks for years.

GEO links are suitable for seamless intermediate connections in a TCP circuit. Nothing is denying the use of many small TCP connections over a broadband link; GEO, fiber, and most broadband Internet connections contain a vast number of separate small pipes. The real issue with *geosynchronous orbit* (GEO) versus *low-Earth orbit* (LEO) is the acceptability of the physical delay for two-way, real-time applications such as telephony or video conferencing. Even then, the physical delay of GEO is balanced somewhat by the increased switching times through a packet-based LEO network. Having multiple narrow TCP pipes across satellite works well. Wide pipes with large buffer sizes can suffer from the high *bit error rate* (BER) of satellites.

For TCP, implementing Selective Acknowledgements (RFC 2018) and fast recovery (RFC 2001) also improves performance in the face of errors. Much work is in progress with the *IP-over-satellite* and the *TCP-over-satellite* working groups.

IP-over-satellite's primary advantage is that it can deliver large amounts of bandwidth to underdeveloped areas where fiber does not exist. IP-over-satellite is also subject to a number of adverse conditions that can significantly impact the effective throughput rates and network efficiency. Bit error rates, congestion, queue management, Window size, and buffer status can all have a serious impact on the overall IP-over-satellite performance curves.

Another problem is the asymmetry encountered in back-channel bandwidth, which in many IP-over-satellite systems is only a fraction of the forward-channel bandwidth. Because TCP emerged in advance of IP-over-satellite, TCP often gets mistakenly identified as the source of the problem. It is the operating environment and not TCP alone that is the problem. TCP is not optimized for conditions encountered during satellite transmissions.

The standard GEO round trip hop of 44,600 miles creates problems with the inherent transmission-related delay breaking down the flow of packets between the sender and receiver, especially at very high-transmission speeds. TCP works well as a general-purpose protocol for a congested environment, such as the Internet. However, TCP perceives a satellite delay and bit error as congestion and inherently slows.

Satellite and ATM

Work has also been done on *Automatic Transfer Mode* (ATM)-over-satellite. Standardization of ATM-over-satellite is underway and several organizations are working in parallel. The ATM Forum's *Wireless-ATM* (WATM) group has done some work on ATM-over-satellite. The *Telecommunications Industries of America* (TIA) Communications and Interoperability group (TIA-TR34.1) works on interoperability specifications that facilitate ATM access and ATM network interconnect in both fixed and mobile satellite networks. TIA is collaborating with the ATM-Forum WATM group. The standard development for satellite ATM networks is provided in ATM Forum contribution 98-0828.

Charting the Rules for the Internet

Everyone knows about the Internet. Few, however, understand how it ticks. Very few people know where the standards come from that shape how the Internet works. These rules come from five groups, four of them loosely organized and consisting primarily of volunteers. The groups don't issue standards in the traditional sense, although their conclusions become the regulations by which the Internet operates worldwide. Instead, they agree on how things should be through discussion. The result is a surprisingly effective, if sometimes convoluted, system for dealing with the fast-growing and often unpredictable nature of the Internet. Like the Internet itself, the groups that oversee it have been recent innovations that continue to evolve and restructure.

The Internet links thousands of networks and millions of users who communicate using hundreds of different types of software. It is able to do so because of TCP/IP, the general procedure for accurately exchanging packets of data adopted by ARPANET in the early 1980s. Other networks soon adopted TCP/IP as well, thus paving the way for the global Internet of today. TCP/IP has been updated several times. It continues to undergo modifications, including those designed to speed throughput, especially for long-terrestrial and satellite links.

Tailoring IP Can Accelerate Throughput

Everyone who uses the Internet wants faster connections. The quest for speed has become a big marketing issue for terrestrial and satellite systems alike. However, no one system is inherently the best for every application. Finding the most efficient way to connect is a matter of matching communications needs to the unique characteristics of each option. The hardware involved is an obvious factor in system speed. Nevertheless, the protocols used by computers to talk to each other ultimately control throughput. Internet protocols are constantly being revised, and faster versions are coming that promise quicker data delivery over links of all kinds. Primarily the *Internet Engineering Task Force* (IETF) coordinates the work. The IETF seeks continual improvement in the software that makes the Internet work, and it prefers solutions that benefit terrestrial as well as satellite network links.

The Internet works because all computers using it abide by the same rules, or protocols. The most fundamental of these rules is the *Internet Protocol* (IP). Its primary function is to provide the *datagram* that carries the address to which a computer is sending a message.

Guaranteeing that messages are received accurately over the Internet is the primary job of the TCP. TCP sends data in segments and waits for a confirmation message from the receiving computer before sending more. It reacts to traffic loads on the network, dynamically regulating the allowable rate of data transfer between the two computers in an attempt to maximize data flow without overwhelming relay points or the capacity of the receiving computer.

TCP facilitates sending large data files accurately over transmission routes that may be plagued with bit errors or other forms of poor link quality. It also reduces throughput when it doesn't receive acknowledgment as quickly as it expects. TCP assumes that any delays are due to traffic congestion within the network. It responds by cutting back on the transmission rate, and then slowly speeds up again as long as no further delays are detected. The idea behind its slow-start mechanism is to match the output of the sending computer to the maximum throughput allowed by the network at any moment. For the most part, TCP works just fine. However, pushing the ultimate throughput capabilities of TCP has revealed shortcomings, especially on high-capacity links with high latency. The problem: high latency slows TCP's slow-start mechanism dramatically, causing the protocol to restrict the amount of data that can be in flight between the sending and receiving computers. This prevents the link from being fully utilized.

The problem affects any high-delay path, whether terrestrial or by satellite. For high-bandwidth links, the result is a sharp drop in efficiency. For example, a message controlled by the most commonly used form of TCP can move through a cross-continent OC-3 fiber link at only about 1.1 Mbps, even though the line can handle 155 Mbps of throughput. A packet of data can propagate across the United States and back through a dedicated copper wire or fiber in about 60 milliseconds. It takes eight to nine times as long—(500 milliseconds) for the same packet to make the round trip via satellite.

Individual LEO spacecraft promises very short propagation delays—as little as 12 milliseconds from ground to satellite and back—because they fly closer to the ground. However, a single LEO spacecraft can't see from coast to coast, so several such birds are required to relay a packet across the United States. LEO birds also are in motion relative to the ground, so their packet latency varies continually, a characteristic that forces current forms of TCP to reassess the allowable throughput rate constantly. The net result is cross-country transmission delays for LEO systems akin to those of long-terrestrial links.

Signal latency isn't the only factor in selecting an appropriate transmission medium, however. Each alternative has its own attractions and challenges. Continuing adjustments to TCP promise to improve performance on any high-latency network link. From a broadband wireless solution, the use of TCP/IP over the Internet, and wireless links the two hottest topics together. Satellite transmission just requires different techniques to support the high-speed communications. The latency and the lack of reliability of the Internet are due primarily to its decentralized nature. Remember that the Internet is a loosely coupled group of networks making a best attempt to deliver. If the delivery is not reliable, then retransmissions are constant. This constant retransmission is at the expense of throughput. Typically, whenever we download any data (or Web site information) the packet will traverse between 17 and 20 hops each with its own buffer and delays. Using satellite transmission rather than landline or terrestrial-based communications can reduce the number of hops and the resultant delay. Satellites in a GEO orbit can add a tremendous amount of delay. Therefore, the use of a LEO can reduce the round-trip delay and improve overall performance. One of the possibilities to satisfy the broadband communications is the network proposed by Teledesic.

Teledesic Technology Overview

The Teledesic network is a high-capacity broadband network that combines the global coverage and low latency of a LEO constellation of satellites, the flexibility and robustness of the Internet, and fiber-like *Quality of Service* (QOS). Termed as an Internet-in-the-Sky™ system, the Teledesic network promises to bring affordable access to interactive broadband communication globally, including to areas that could not be served economically by other means such as wires or terrestrial-based cellular systems.

The Teledesic network can serve as the access link between a user and a gateway into a terrestrial network, or as the means to link users or networks together and support millions of simultaneous users when full deployment is complete.

Seamless Compatibility

Geostationary satellite (GEO) systems require changes to terrestrial network standards and protocols to accommodate their inherent high latency. Teledesic's objective is to meet current network standards rather than to change them. This requires a systems design that has low latency, low bit error rates; high-service availability; and flexible, broadband capacity.

The Teledesic Network

The network consists of a ground segment (terminals, network gateways and network operations, and control systems) and a space segment (the satellite-based switch network that provides the communication links among terminals). Terminals are the edge of the network and interface both between the satellite network and the terrestrial end users. These terminals perform the translation between the Teledesic internal protocols and the standard protocols, thereby isolating the satellite-based core network from complexity and change, as shown in Figure 9-4.

Teledesic terminals communicate directly with the satellite network and support many data rates. The terminals also support standard network protocols, including IP, ISDN, ATM, and others. Although optimized for service to fixed-site terminals, the network is capable of serving mobile terminals, such as those for maritime and aviation applications. Most users will have two-way connections that provide up to 64 Mbps on the downlink and up to 2 Mbps on the uplink. Broadband terminals will offer 64 Mbps of two-way capacity.

The capability to handle multiple channel rates, protocols, and service priorities provides the flexibility to support a wide range of applications including the Internet, corporate intranets, multimedia communication, LAN interconnect, wireless backhaul, and small office/home office applications.

Terminals also provide the interconnection points for the network's *Constellation Operations Control Centers* (COCC) and *Network Operations*

Figure 9-4

The network interfaces (Source: Teledesic)

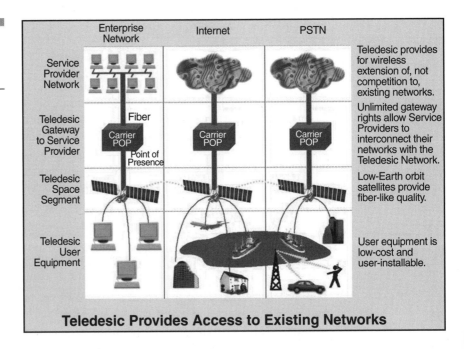

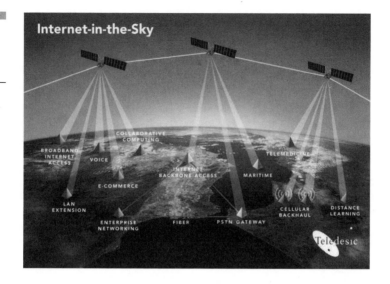

Figure 9-5
The Teledesic
network
(Source: Teledesic)

Control Centers (NOCC). COCCs coordinate the initial deployment of the satellites, replenishment of spares, fault diagnosis, repair, and de-orbiting. Figure 9-5 illustrates the network.

Fast Packet Switching

Teledesic's space-based network uses fast packet switching. Communications are treated within the network as streams of short, fixed-length packets. Each packet contains a header that includes the destination address and sequence information, an error-control section used to verify the integrity of the header, and a payload section that carries the digitally encoded user data (voice, video, data, and so on). Conversion to and from the packet format takes place in the terminals at the edge of the network.

The topology of a LEO-based network is dynamic. The network must continually adapt to these changing conditions to achieve the optimal (least-delay) connections between terminals. The Teledesic network uses a combination of destination-based packet addressing and a distributed, adaptive packet-routing algorithm to achieve low delay and low-delay variability across the network. Each packet carries the network address of the destination terminal, and each node independently selects the least-delay route to that destination. Packets of the same session may follow different

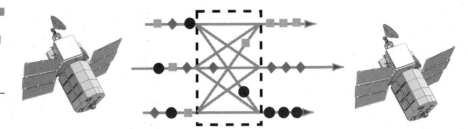

paths through the network routing traffic, as shown in Figure 9-6. The terminal at the destination buffers and if necessary, reorders the received packets to eliminate the effect of timing variations.

The Satellite Constellation

Each satellite is a node in the fast packet-switched network and has intersatellite communication links with other satellites in the same and adjacent orbital planes. This interconnection arrangement forms a robust non-hierarchical mesh, or *geodesic*, network that is tolerant to faults and local congestion. The network combines the advantages of a circuit-switched network (low-delay digital pipes) and a packet-switched network (efficient handling of multirate and bursty data).

From a network viewpoint, a large constellation of interlinked switch nodes offers a number of advantages in terms of service quality, reliability, and capacity. The richly interconnected mesh network is a robust, fault-tolerant design that automatically adapts to topology changes and to congested or faulty nodes and links. To achieve high system capacity and channel density, each satellite is able to concentrate a large amount of capacity in its relatively small coverage area. Overlapping coverage areas plus the use of on-orbit spares permit the rapid repair of the network whenever a satellite failure results in a coverage gap. In essence, the system reliability is built into the constellation as a whole rather than being vulnerable to the failure of a single satellite.

The Ka-band is the lowest frequency band with sufficient spectrum to meet Teledesic's broadband service, quality, and capacity objectives. The terminal-satellite communication links operate within the portion of the Ka-frequency band that has been identified internationally for non-geostationary fixed satellite service and in the U.S., licensed for use by Teledesic. Downlinks operate between 18.8 GHz and 19.3 GHz, and uplinks operate between 28.6 GHz and 29.1 GHz. Communication links at these frequencies are degraded by rain and blocked by obstacles in the line of sight. To avoid obstacles and limit the portion of the path exposed to rain requires that the satellite serving a terminal be at a high-elevation angle above the horizon. The Teledesic constellation ensures a minimum elevation angle (mask angle) of 401/4 within its entire service area. Using this design, the Teledesic network is able to achieve availability of 99.9 percent or greater.

Latency is a critical parameter of communication service quality, particularly for interactive communication and for many standard data protocols. To be compatible with the latency requirements of protocols developed for the terrestrial broadband infrastructure, Teledesic satellites operate at a low altitude, under 1,400 kilometers. The combination of a high-mask angle and LEO result in a relatively small satellite coverage zone, or footprint, that enables efficient spectrum reuse but requires a large number of satellites to serve the entire Earth. In the initial constellation, the Teledesic network will consist of 288 operational satellites, divided into 12 planes, each with 24 satellites.

Multiple Access

Because the network uses wireless access, communication channels are not dedicated to terminals on a permanent basis. The channel resources associated with a cell are shared among terminals in that cell, with capacity assigned on demand to meet their current needs. This flexibility enables Teledesic to efficiently handle a variety of user needs:

- From occasional use to full-time use
- From bursty to constant bit-rate applications
- From low-rate to high-rate data
- From low-usage density areas to areas of relatively high-usage density

A multiple access scheme implemented within the terminals and the satellite serving the cell manages the sharing of channel resources among terminals. Within a cell, channel sharing is accomplished with a combination of *Multifrequency Time Division Multiple Access* (MF-TDMA) on the uplink and *Asynchronous Time Division Multiplexing Access* (ATDMA) on the downlink.

Network Capacity

To make efficient use of the radio spectrum, frequencies are allocated dynamically and reused many times within each satellite footprint. The Teledesic network supports bandwidth-on-demand, enabling a user to request and release capacity as needed. This enables users to pay only for the capacity they actually use, and for the network to support a much higher number of users. Thus, the Teledesic network is designed to support millions of simultaneous users. The network scales gracefully to much higher capacity by adding additional satellites, which are shown in Figure 9-7.

Teledesic intends to provide high-quality, broadband communications services to corporations, governments, non-governmental organizations, small businesses, and communities throughout the world. Through the use of satellite technology, the network will offer high-bandwidth connectivity to all corners of the globe.

Virtually all global markets for telecommunications services are experiencing remarkable growth. Electronic commerce, e-business, and basic connectivity are exploding, causing huge increases in the need for advanced telecommunications infrastructure. This growth will accelerate in the decades to come as expanding user populations and increasingly sophisticated applications enhance the world's appetite for bandwidth. Teledesic

Figure 9-7
The Teledesic satellites
(Source: Teledesic)

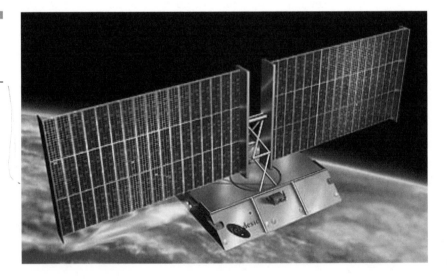

will help satisfy this appetite by building the world's first truly global broadband data network.

Wireless Local Loop (WLL)

Sometimes called *radio in the loop* (RITL) or *fixed-radio access* (FRA), *wireless local loop* (WLL) is a system that connects subscribers to the *public-switched telephone network* (PSTN) using radio signals as a substitute for copper for all or part of the connection between the subscriber and the switch. This includes cordless access systems, proprietary FRA, and fixed-cellular systems.

Since the advent of the telephone system, copper wire has traditionally provided the link in the local loop between the telephone subscriber and the local exchange. However, we keep hearing that copper's heyday in the local loop is ending. Economic imperatives and new technologies open the door for WLL solutions. WLL uses wireless technology coupled with line interfaces and other circuitry to complete the last mile between the customer premise and the exchange equipment. According to the research firm MTA-EMCI, the worldwide WLL market will reach 202 million subscribers by the year 2005. Many research estimates indicate that 172 million to 307 million subscribers will have a demand for WLL service. Actual service will be provided for 140 to 150 million subscribers by the year 2004, as shown in Figure 9-8.

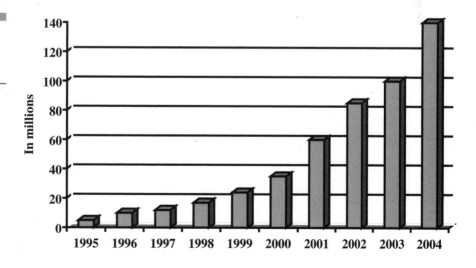

Figure 9-8
Demand for broadband wireless local loop

WLL Technology Shakeout

The WLL revolution is underway. WLL suppliers and operators are flocking to emerging markets, using whatever available wireless and line interface technologies at hand to achieve fast time to market. Because no definitive WLL standards are available, vendors still face a bewildering choice of fixed-access, mobile, and digital cordless technologies. The appropriate technology will depend on an array of considerations, such as size and population density of the geographic area (rural versus urban) and the service needs of the subscriber base (residential versus business, POTS versus data access). In fact, different wireless technologies will serve some applications better than others for many reasons. The challenge for WLL vendors is to identify the optimal wireless protocol for their unique application needs, reduce cost per subscriber through silicon, and deliver integrated solutions to the marketplace.

Architecture of a Wireless Downstream System

Engineering of a wireless system is complicated by several factors:

- The physical wireless transmitter location may be different from the Internet head end location.
- Line-of-sight transmission is required.
- The signal power falls proportionately with distance.
- Different fixed transmission frequencies are used.
- Multipath distortion can occur.
- A receiving antenna is required on the building.
- Two-way operation is limited.

Some of these requirements interact in the following ways:

- If one had complete freedom, the transmitter would be at a convenient site for an Internet service connection. However, we must have line-of-sight transmission to customers. These two requirements conflict about half the time so the transmitter and the Internet head end cannot be co-located.
- The site for the wireless antenna and the transmitter is usually chosen to provide line-of-sight television to a metropolitan area or several

rural communities. The frequency bands were licensed for this wide area coverage. Therefore, the transmitter site may be physically inaccessible on a mountain.

■ The nearest point of connection to the Internet is often downtown. This is often the best place for the Internet head end maintenance service connection yet has line-of-sight transmission to customers.

As shown in Figure 9-9, the *Multichannel Multipoint Distribution Service* (MMDS) transmitter may not cover the whole area. A low-power booster may be needed to fill in areas where line-of-sight does not exist.

Frequency Bands and Limitations

The frequency band determines the type of receiving antenna and the coverage, or range, of the transmitter. MMDS, *Instructional Television Fixed Service* (ITFS), and *Multipoint Distribution Service* (MDS) analog television transmitters were the first used for Internet access. The bands were underutilized for television and one, two, or even three digital 2-MHz subchannels could be made to work with adjustments to the analog transmitter diplexers and filters. Digital transmitters have recently been made available from several vendors. WCS is a newly available band with wireless transmitters and down converters under development. *Ultra High Frequency* (UHF) and *Very High Frequency* (VHF) low-power transmitters are in service for Internet downstream access using experimental licenses. Most bandwidths are 6 MHz downstream, whereas WCS is 5 MHz. Table 9-2 lists the possible services for the broadband local loop. Table 9-3 lists the frequency bands available for these services. The primary services that have been selected have been *local multipoint distribution services* (LMDS) and MMDS.

Receiving the Signal at the Subscriber

The frequency band determines the type of receiving antenna and down converter (if needed). LPTV signals are received by the cable modem using a normal outside TV antenna, possibly with an amplifier but paying more

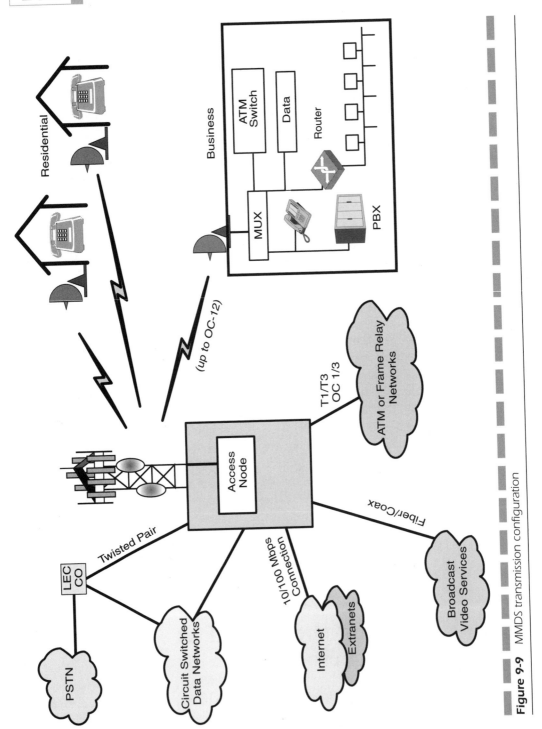

Figure 9-9 MMDS transmission configuration

Table 9-2

List of Possible Services for Broadband Distribution

Name	Description and Basic Usage
MMDS	Multichannel Multipoint Distribution Service—Most of these TV transmitters are analog and require upgrade to digital unless subchannelization is used.
MDS Multipoint Distribution Service	Analog TV or not used.
WCS Wireless Communications Services	New.
ITFS Instructional Television Fixed Service	Educational service includes Internet access.
LMDS Local Multipoint Distribution Service	New.
ISM Instructional Scientific and Medical	Unlicensed bands used for LANs and for the return path of two-way modem systems.

attention to the signal level than one would with an analog television set. The level must be close to the high end of the range acceptable to the modem as the signal is more likely to fade than increase in level.

MMDS, MDS, and WCS require a small antenna integrated with a down converter mounted on the roof or on the side of the building. These low-cost units are one-third the price of the cable modem. The following are points to note in a wireless environment:

■ Locate the antenna to minimize multipath.

■ Make sure the antenna is located to receive the direct signal rather than a stronger reflected signal off a building, which may vary in level.

■ Work with the down converter vendor to achieve the simplest selection of amplifier gain and antenna gain for the coverage area to minimize the number of different models needed.

■ Use the appropriate path loss calculations for microwave links. The calculation for multiple modulators through a common transmitter (subchannelization) is a little different from a single signal on a transmitter.

■ A down converter installation engineered for a cable modem can also support video service. The reverse may not be true as the cable modem has specific level requirements.

Table 9-3

Frequency Bands

Table 9-3

Frequency Bands

Name	Frequency MHz	Notes
MMDS	2,500–2,686	Thirty-one individual 6-MHz television channels (transmitters) including ITFS. Some operators have only four channels. It ranges to 35 miles, requires line-of-sight, and is affected by multipath.
MDS1	2,150–2,156	Single channel 6 MHz. See MMDS.
MDS2	2,156–2,162	Single channel 6 MHz. See MMDS.
MDS2A	2,156–2,160	MDS2 truncated on one side to 4 MHz.
WCS	2,305–2,320	5-MHz or 10-MHz blocks.
WCS	2,345–2,360	5-MHz or 10-MHz blocks (new in May 1997).
ITFS	2,500–2,690	6-MHz channels shared with MMDS.
LPTV	54–72 78–88 174–216 470–806	Low-power broadcast, 6-MHz channels, experimental licenses Low power can be 50-Kwatts effective radiated power. (This 174–216 includes the antenna gain.) Line-of-sight operation is advised.
LMDS	27,500–28,350 31,000–31,300	Short range, 3 miles, 20-MHz channels (new). Propagation is affected by rain.
ISM	902–928 2,400–2,483.5	Short range, 0.5 miles spread spectrum omni-directional. Short range, similar to 900 MHz but can also be engineered beyond 15 miles point-to-point as the return path for a cable modem system.

The cable modem tuner/demodulator will accept a limited variation in input level. Depending on manufacturer, model, frequency of operation, and MMDS environment, the path loss is stable except for multipath distortion, and the antenna or down converter gain can be selected to provide the right level to the cable modem. In the case of LPTV, path fading is more common, but an outdoor antenna with line-of-sight operation will minimize the level variations from multipath.

Although many of these discussions center on new technology, the two that have gained the most momentum are the concept of the wireless local loop (WLL) and the local multipoint distribution services (LMDS). The following section will consider these concepts more closely.

Wireless Local Loop (WLL)

The industry in general has placed a lot of emphasis on the WLL and predicts that millions of subscribers will enjoy the benefits of an untethered communications before the turn of the century. This may or may not be aggressive, but it signals the point that the competitive machine is in full swing at the last mile. Much of the growth being discussed will occur in areas where an infrastructure does not exist, such as third-world countries installing the initial communications systems to the residential and business user. Many countries across the globe still do not have basic *plain old telephone service* (POTS), so it makes sense to consider a wireless connection. In some cases the use of a *radio in the loop* (RITL) concept or a *fixed wireless radio access* (FWRA) concept are what the countries have dubbed the services. Countries like Brazil and China will reap many benefits from using a WLL concept, both financially and in the speed of installation. The cost of installation on a per-user basis is much more favorable. One set of statistics shows the difference of the installed cable versus a wireless local access method, as shown in Table 9-4.

However, the emerging underdeveloped countries are not the only places where WLL technology will be used. Instead, the developed countries around the world may also take advantage of the economies of scale and the financial benefits of installing the wireless local access. As a result, as many as 50 million access lines may be deployed worldwide shortly after the turn of the century and rapid growth may follow the initial installations. The day of installing copper to the door has ceased; instead, the wireless technologies may be the mode of choice for the future. No longer can the carriers afford the cost of installation and maintenance for the copper local loop.

Table 9-4

Cost Comparisons for Wired versus Wireless Local Loop

Technology Used	Cost per User[1]
Copper local loop	$ 5,500.00
Wireless Local Loop	$500.00–$800.00[2]

[1] Cost is considered in U.S. dollars.

[2] This figure will drop rapidly to approximately $200–$300 per user as deployment continues and economies of scale are achieved.

Figure 9-10 is a representation of the overall concept of the WLL. This does not use specific technology but serves as a model for the carriers considering the use of wireless technology.

Not for Everyone

The wireless local loop will see many new opportunists jump into the market; however, many will not survive. Either the providers will be underfunded and will not survive the competition, or the larger providers looking for market share in an area of operation will *gobble* the smaller local providers up. In either case, the number of providers will change and the operators will continually be looking for new and competitive approaches to attract customers. Full-service providers will offer the list of services, as shown in Table 9-5. Others may offer pieces of these services. The point is that the end user is looking for a one-stop shopping approach and the leverage that comes with bundled services.

Too many providers will jeopardize the success of many, causing some form of shakeout, but one must consider the end user is willing to use one or more of the providers listed in Table 9-5. A merger or a joint offering will likely occur from partnering providers to get to the consumer's door. When one looks at these offerings and carriers, it is obvious that the services are disjointed. Some providers offer all the services, whereas others are just planning the possible services they will offer. However, the infrastructure they choose to install may have an impact on their capability to service future demands. No one answer or solution jumps out right now, but changes will occur rather quickly in this business. The economics of getting the consumer (both residential and business) to buy into more than one offering will set the stage for future services. One can add the numbers and see where the providers want to take this. Table 9-6 shows a summary of service offerings (on average) for the services used by the consumer. In this particular scenario, the consumer is a home office-based user or a residential user whose needs include various bundled services. If a carrier can offer the bundled services for a moderate decrease in the monthly costs, one can expect 65 percent of those approached to churn.

Using these bundled one-stop pricing models, one can expect that the residential and small business customer will be tempted to use the service provider. Note that not all service providers will offer the equipment (such as the PC or the modems). However, some of the pieces may not be required. For example, many of the WLL providers include the analog cellular suppliers and the PCS suppliers whose recent advertisements state that

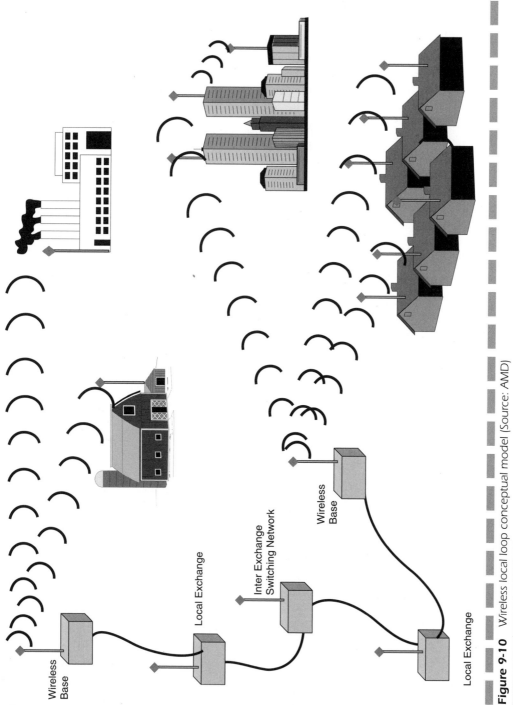

Figure 9-10 Wireless local loop conceptual model (Source: AMD)

Table 9-5

Summary of Service
Offerings and
Providers Today

						Internet Access (High Speed)	
Provider	Voice	Data Low Speed	Point-to-Point Data	CATV	Video Conferencing		Multimedia Services
Cable Companies	Yes	Not avail., but possible	Not avail., but possible	Yes	Not avail., but possible	Yes, 10 Mbps	Not avail.
LECs	Yes	Yes	Yes	No, but planned	Yes	Yes, 1.5 Mbps	Limited
CLECs	Yes	Yes	Yes	No, but planned	Yes, Limited	Yes,1.5 Mbps	Limited
IECs	Yes	Yes	Yes, but not local loop	No, merge with CATV	Yes, Limited	Yes, 1.5 Mbps	Limited
WLL	Yes	Yes	Limited	No, but possible	Limited	Yes, 10 Mbps	Possible
Cellular providers	Yes	Yes	No	No	No	No	No
PCS providers	Yes	Yes	No	No	No	Limited	No

The table header "Services Offered by Carriers Today and Future Offerings" spans the data columns.

consumers can remove their wired telephone and use the cellular or PCS service for their home and their business needs. This is possible and has some merit. Thus, if the consumer takes this provider up on this advertisement, the carrier loses the bundling of a $25.00 monthly dial tone service. However, the cost of the cellular plan will increase in the number of minutes used, driving that plan cost up higher. This may become an even trade.

Another point is the cost of the infrastructure. Once the CATV companies have delivered the basic cable services, for example, the cost of any added usage or shared bandwidth on their infrastructure is usually marginal. Thus, the profitability and mark-up is that much higher. The wired carriers understand the benefit of the one-stop shopping, and now the WLL carriers are learning very quickly.

Table 9-6

Bundled vs. Individual Services Plans

Service Offering	Average Monthly Price	Bundled Price
CATV (basic cable)	$ 8.00	$100.00
Extended channel services[3]	$ 23.00	
Basic Internet Service Provider access	$ 20.00	
Data access for Internet at dial-up rates	$ 25.00	
Dial tone for voice	$ 25.00	
High-speed Internet access at 1+ Mbps	$ 50.00	$ 35.00
Long-distance services typical customer	$ 25.00	$ 15.00
Equipment costs amortized (PC, modem, telephone, etc.) monthly	$ 30.00	$ 30.00
Cellular phone basic plan	$ 50.00	$ 30.00
Total monthly fees	$256.00	$210.00[4]

[3] Excludes premium channel services (HBO, Showtime, and so on) and pay-per-view services.

[4] The goal of the providers with one-stop shopping is to offer all the services at approximately $200.00 per month.

What of the Bandwidth?

The bandwidth necessary for each of these services changes the rules considerably. Many of the WLL providers' backbones do not have enough bandwidth to support the number of users and the high-speed services. For this reason, the marriage of the providers may occur sooner than expected. If a cellular provider joins forces with a WLL microwave supplier, then the bandwidth for the fixed needs at the door is ensured while the cellular provider handles the demands of the roaming user. These combinations and permutations can be very complicated as the number of providers expands and the services they offer shift in any direction. The interesting point will be to see how the total market plays out with an expectation that approximately five to seven providers will dominate and the rest will be absorbed or fail.

Enter Local Multipoint Distribution Services (LMDS)

Local Multipoint Distribution Services (LMDS), as its name implies, is a broadband wireless technology used to deliver the multiple service offerings in a localized area. The services possible with LMDS include the following:

- Voice dial-up services
- Data
- Internet access
- Video

Just as the network providers were getting used to the battlegrounds between the *Incumbent LECs* (ILECs) and the new providers, RF spectrum was freed up around the world to support access and bypass services. Typically, the services operate in the radio frequency spectrum above 25 GHz, depending on the licenses and spectrum controlled by the regulatory bodies. This offering operates as a point-to-point broadband wireless access method, which can provide two-way services. Because LMDS operates in the higher frequencies, the radio signals are limited to approximately 5 miles of point-to-point service. This makes it somewhat like a cellular operation in the way the carriers lay out their operations and cells. An architectural concept for the LMDS operation is shown in Figure 9-11 from the perspective of the supplier to the user. This figure uses some of the premise that the service is constrained to a localized area.[5]

The Reasoning Behind LMDS

Point-to-point fixed microwave radio has been in use for decades in the local loop environment. Many organizations (individual businesses, utility companies, and so on) required dedicated access to their own private network facilities or to a carrier's *Point of Presence* (POP). As they would approach the ILEC, the cost to run high-speed services to the business consumer's

---

[5] Occasionally in uncongested and unpopulated areas, the signals are transmitted in much wider areas of coverage, similar to other wireless technologies. This reference to the 5 miles is within populated areas and the obstacles that will be encountered within the areas.

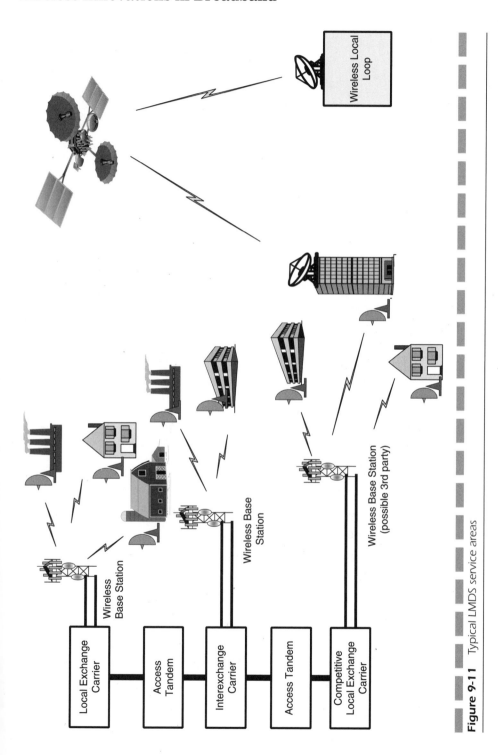

Figure 9-11 Typical LMDS service areas

Table 9-7

Typical Microwave
Distances, Bands,
and Operations

Frequency Band	Distances	Use
2–6 GHz	30 miles	Commercial, utility, fixed operation, TV
10–12 GHz	20 miles	Commercial, utility, fixed operation, TV, DBS
18 GHz	7 miles	Business, limited fixed operation
23 GHz	5 miles	Business, limited fixed operation
25 GHz and above	3–5 miles	Business, bypass operation

door was typically prohibitive. The monopoly owning the embedded infrastructure could demand any price that seemed appropriate. This was met with objections from the user, but as long as the monopoly existed, few choices were available. The businesses, therefore, demanded frequency spectrum to install their own infrastructure at the last mile. The connection was typically in an especially set-aside frequency band using distance sensitive frequencies, as listed in Table 9-7.

The fixed access methods using microwave in the past included the following problems:

■ Local ordinances were unfavorable to the use of the technology.

■ Local regulatory bodies had several restrictions.

■ Federal authorities limited the use.

■ The cost of building the tower was high.

■ The cost of security for the site was high.

■ Local power and other utilities were not readily available.

■ The cost of the equipment was very high.

■ The maintenance costs were unnecessarily high.

■ FCC-licensed technicians were required to do the maintenance.[6]

■ Line-of-sight frequencies were not readily available.

Each of these areas was somewhat limiting to the demands of the end user's ability to get access to the fixed point-to-point microwave systems. The largest organizations could financially justify the use of this service

---

[6] Reference to the U.S.-based operations—in other countries, similar requirements prevailed (such as CRTC in Canada and PTT in other parts of the world.)

because their needs were more demanding. However, smaller organizations had to rely on alternative methods or service bureaus that could provide the access at a reduced rate.

From a carrier's perspective, however, the equation changes very quickly. By using LMDS services, a new provider can install the systems more readily due to the competitive environment being introduced worldwide. The monopolies no longer mandate or dictate what the local connectivity will be like. The new providers can achieve the benefits of the LMDS world through

- Lower cost entry into the market
- Costs are deferred to later when services are needed. This moves the pricing model from fixed to variable costs associated with demand as opposed to fixed size increments.
- Return on investments are achieved more quickly, encouraging the provider to enter the market.
- Less risk of customer churn leaving the carrier stuck with large investments
- Ease of installation and licensing makes the implementation faster
- Standards-based services and equipment, minimizing obsolescence, and proprietary solutions

The carriers seem to have found a nirvana of technology and financial benefit in a single solution. The real issues then begin to work around the need, demand, and the method of delivery. Not all systems are implemented exactly the same, so the carrier still has some choices enabling even greater flexibility in delivering the bandwidth to the door.

Network Architectures Available to the Carriers

Various means of installing gives the carrier choices as already stated. The bulk of the carriers will likely standardize on a straight point-to-point connectivity solution for their customers. Point-to-point TV distribution can also be provided with the LMDS offering. This increases the attractiveness of the LMDS supplier when adding to the TV distribution capability the other services desirable by the end user such as voice, data (IP), and multimedia applications. The architecture of the LMDS will lend itself to these point-to-point services nicely. The primary pieces constituting the LMDS system are as follows:

- **Network Operations Center (NOC)**   Contains all the management functions that manage all the components of a much larger infrastructure.

- **Cabled infrastructure**   Usually fiber-based to connect the components of the LMDS to the public-switched and private networks. The cabling will consist of T1/T3 or OC-1, OC-3, or OC-12 connecting to the ATM and Internet backbones.

- **Base station**   is where the fiber-to-radio frequency conversion takes place. The modulation of the signal across the airwaves also occurs here.

- **Customer equipment**   Can vary from user to user and by vendor to satisfy the demands of the consumer.

The architecture also varies in the modulation of the signals onto the RF (airwaves) based on the chosen strategy of the vendor. Two predominant methods of using the technology are to use an analog interface such as *Frequency Division Multiple Access* (FDMA) or a digital interface using *Time Division Multiple Access* (TDMA). The choice will vary depending on the density of the sectors being served, the available financial choices, and the overall quality desired. The more common implementation is to use an FDMA technique to serve the customer. The use of FDMA provides better coverage and density of the applications being served and it uses a higher modulation technique to satisfy the system demands.

TCP/IP over LMDS

Radio-based transmission for an IP network is particularly attractive to new entrants in the telecommunications marketplace. Because these providers do not have a wired infrastructure and because they will have to negotiate with the incumbent LEC to gain that access, the providers see LMDS and WLL as being their initial salvation. Moreover, the wireless infrastructure enables the provider to support both fixed-location and mobile IP users simultaneously. The main disadvantages include weather degradation and line-of-sight requirements.

To be competitive, the broadband access radio network requirements are

- Low modem costs, comparable of less than copper cost
- Total sharing capacity of up to several megabits per second per radio cell (bandwidth on demand)
- Bandwidth efficiency that will allow maximum users for data transmission
- Highly flexible to support future demands (voice, data, video, and multimedia) in a customizable format on a subscriber basis
- Flexible networking standards enabling increased bandwidth or reconfiguration on-the-fly

Most of the in-place systems are based on point-to-point networking architectures. Each radio system has its own capacity using multiples of 2-Mbps throughput. As the industry matures, the implementations for the future will be ATM-based WLL and IP-centric networking standards to support the variations of user input. The LMDS systems today deliver this form of bandwidth for data transmission. MMDS was primarily developed to deliver CATV services instead of the interactive multimedia services. Thus, LMDS is better suited to deliver broadband data over wireless communications. The IEEE standards committees have adopted the 802.16 standard, which includes LMDS among other systems. These systems operate in the 30-GHz frequency band and support speeds for small- to medium-sized organizations ranging from 2 to 155 Mbps. To support the standard TCP/IP architectures, however, aggressive coding and transmission of LMDS services enables us to achieve better tuning. One must be aware of the consequences of using TCP over fixed wireless networks and the need to maintain packet loss and reliability standards or else the TCP networks become virtually useless. The unreliable networking standards of wireless can be overcome with an LMDS radio-based system.

The use of broadband wireless systems can support speeds of 2 to 64 Mbps in a satellite network, coupled with the LMDS architectures to gain upwards of 155 Mbps for today. These systems cost on average $30 to $60K per end but are rapidly declining as more usage takes place. As we progress into the future, speeds in excess of 622 Mbps will be readily available across wireless systems using free space optics and infrared light-based systems. The optical replacements are bringing the costs of access to price parity with wireline services. An OC-12 (622-Mbps) access will be less expensive to install than a fiber link. The cost of fiber can be $100,000 to $250,000 per mile to install (including all trenching and permits), whereas the cost of a radio/optic system will be approximately $30 to $40K per mile equivalent.

Moreover, the optical replacements will support speeds today (in the labs) up to 2.4 Gbps and will steadily climb to the 10-Gbps range in the next two to three years. These will be more economical as time progresses, but today they offer much hope to the new providers who are not interested in building a wireline infrastructure. Applications for these speeds are not yet capable of using the capacities of this form of throughput. However, the future multimedia networks will begin to stress capacities and services. This is only the beginning!

10

Emerging Wireless Standards

Wireless Standards

Wireless continues to develop around the world. Several different standards committees are working on integrating wireless architecture into the overall fold of the network.

Get ready! As the convergence of wireless technology and the Internet continue at an escalating pace, the new possibilities created by 3G and 4G technologies appear endless. Preparing for the revolution, existing TDMA operators must evolve their networks to take advantage of Mobile Multimedia applications and the eventual shift to an all-IP architecture. One way to do that is through the evolution of *General Packet Radio Services* (GPRS). However, soon after we see the installation of GPRS, some operators will begin the next step in the evolution process to *Enhanced Data for Global Environment* (EDGE). With EDGE, existing TDMA networks can host a variety of new applications, including the following:

- Online e-mail
- Access to the World Wide Web
- Enhanced short message services
- Wireless imaging with instant photos or graphics
- Video services
- Document/information sharing
- Surveillance
- Voice messaging via Internet
- Broadcasting

At the same time, some operators will skip the step to EDGE and go directly to *Universal Mobile Telecommunications Services* (UMTS), or what we consider to be a *third-generation* (3G) technology. The steps are shown in Figure 10-1, as the carriers choose which way to proceed.

Using a timing window, the evolution of wireless to 3G systems are shown in Figure 10-2, using the evolution of the various techniques that emerged over the years.

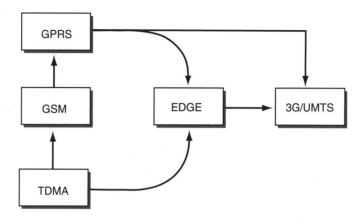

Figure 10-1
The evolution to
UMTS choices

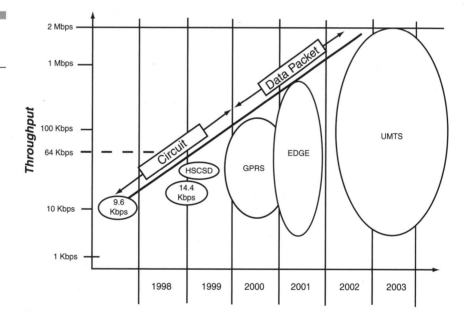

Figure 10-2
Time line for
3G/UMTS

GPRS

Probably the most important aspects of GPRS are that it enables data transmission speeds up to 170 Kbps; it is packet based, and it supports the leading data communications protocols (IP and X. 25).

GPRS operates at much higher speeds than current networks, providing advantages from a software perspective. Wireless middleware currently is required to enable slow speed mobile clients to work with fast networks for applications such as e-mail, databases, groupware, or Internet access. With GPRS, wireless middleware will probably be unnecessary, making it easier to deploy wireless solutions.

Although current wireless applications are text oriented, GPRS' high throughput finally makes multimedia content, including graphics, voice, and video practical. Imagine participating in a videoconference while waiting for your flight at the airport (something that is completely out of the question with today's data networks). The following illustration shows the use of the hand held devices for checking schedules at high-speed connections.

Why is packet data technology important? Because packet networks provide a seamless and immediate connection to the Internet or corporate intranet, enabling access to existing Internet applications, such as e-mail and Web browsing, without dialing into an ISP. The advantage of a packet-based approach is that GPRS only uses the medium, in this case the radio link, only as long as data is being sent or received. Multiple users can share the same radio channel very efficiently. In contrast, with current circuit-switched connections, users have dedicated connections during their entire call, even if they are not sending data. Many applications have idle periods during a session. With packet data, users will only pay for the amount of data they actually communicate, and not the idle time. In fact, with GPRS, users could be "virtually" connected for hours at a time and only incur modest connect charges.

Although packet-based communications works well with all types of communications, it is especially well suited for frequent transmission of small amounts of data. We refer to this as short and bursty, such as "real time" e-mail and dispatch (vehicles and field service). Packet is equally well suited for large batch operations and other applications involving large file transfers. However, when using large file transfers, the cost can become very expensive compared to circuit switched data transmissions. GPRS supports the *Internet Protocol* (IP) as well as the X.25 protocol. IP support is increasingly more important as companies look to the Internet as a way for their remote workers to access corporate intranets. This is true when using a VPN. In the case of VPNs, GPRS works well because of its *GPRS Tunneling Protocol* (GTP) that can secure the mobile data while in transit on the wireless networks, and IPsec transfers can be used when transiting the wireline networks. The GTP is shown in Figure 10-3.

The IP is ubiquitous and familiar, but what is X.25, and why is support for it important? X.25 defines a set of communications protocols that, prior to the Internet, constituted the basis of the world's largest packet data networks. These X.25 networks are still widely used, especially in Europe and the Far East. Wireless access to these networks will benefit many organizations. Any existing IP or X.25 application will now be able to operate over a GSM cellular connection. Think of cellular networks with GPRS service as wireless extensions of the Internet and existing X.25 networks similar to a *local area network* (LAN) connection. As a LAN connection, once a GPRS mobile station registers with the network, it is ready to send and receive packets. A user with a laptop computer could be working on a document without even thinking about being connected, and then automatically receive new e-mail—although this is not 100 percent available today, it is

Figure 10-3
GTP with VPNs

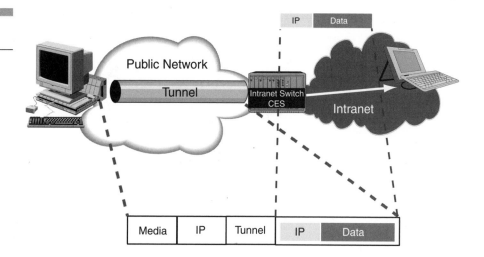

coming. The user may decide to continue working on the document, then half an hour later, read the e-mail message and reply to it. Throughout this time, the user has had a network connection yet never dialed in. Furthermore, some versions of GPRS terminals enable for simultaneous voice and data communication. The user can receive incoming calls or make outgoing calls while in a data session.

Because there is almost minimal delay before sending data, GPRS is ideally suited for applications such as

- Extended communications sessions
- E-mail communications
- Chat
- Database queries
- Dispatch
- Stock updates

In addition, GPRS will remove many of the obstacles from the use of multimedia, graphical Web-based applications because of the higher throughput that is possible. Mobile users will easily use graphically intensive Web-based applications (Map Quest) to obtain directions. The protocol stack supports a variety of interfaces and links multiple networks together as shown in Figure 10-4.

Because GPRS supports standard networking protocols, configuring computers to work with GPRS will be very straightforward. In the case of

Figure 10-4
GPRS protocol stack

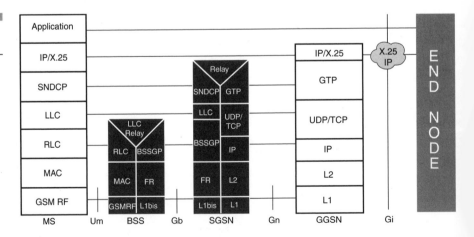

IP communications, one can use existing TCP/IP protocol stacks. TCP/IP stacks are readily available for most other platforms as well. With all the developments in the handheld computer area, expect a multitude of hardware platforms to take advantage of GPRS:

■ Laptops or handheld computers connected to GPRS-capable cell phones or external modems, as shown in illustration 2.

■ Laptops or handheld computer with GPRS-capable PC Card modems, as shown in illustration 3 using a PCMCIA card with an external antenna.

■ Smart phones that have full screen capability.

■ Cell phones employing micro-browsers using the *Wireless Application Protocol* (WAP).

■ Dedicated equipment with integrated GPRS capability, for example, mobile credit-card swipes.

GPRS coincides with another important technology development: the replacement of a cable connection to a cell phone by a short radio link called "Bluetooth."

EDGE

Beyond GPRS, EDGE takes the cellular community one step closer to UMTS. It provides higher data rates than GPRS and introduces a new modulation scheme called 8-PSK (Phase Shift Keying). The TDMA community also adopted EDGE for their migration to UMTS. The data rates allocated for EDGE start at 384 Kbps and above as a second stage to GPRS. EDGE uses the same modulation techniques as many of our existing TDMA infrastructures using *Gaussian Minimum Shift Keying* (GMSK) 8-PSK. Moreover, EDGE uses a combination of FDMA and TDMA as the multiple access control methods. If we look at this from an OSI stack model, EDGE uses FDMA and TDMA at the MAC layer (bottom half of layer 2 OSI). The protocol stack for EDGE is shown in Figure 10-5.

The channels separations are 45 MHz and the carrier spacing is a 200 kHz channel capacity, the same as GSM and GPRS. The number of TDMA slots on each carrier is the same (8) as the GSM and GPRS architecture. When a mobile station wishes to transmit its data, it can request and use from one to eight time slots per TDMA frame. Connectivity is handled via a packet switched data network such as IP and X.25. These can be public data networks or private data networks.

Figure 10-5
EDGE protocol stack

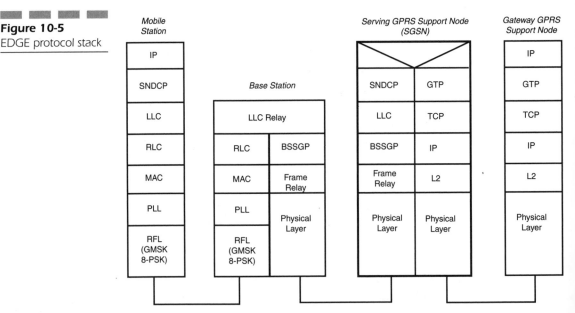

Although most carriers and service providers have plans to deploy enhanced mobile wireless services at higher speeds, the rollout of high-bandwidth wireless transport technology still faces many possibilities. On a positive note, widespread demand will be sufficient enough to support cellular enhancements like high-speed data services and expanded voice capacity. Competitive pressures will also compel service providers to upgrade. The ITU-R has actually established five different standards that fall into the category of 3G/UMTS. Moreover, the telecommunications industry is growing increasingly impatient to test the world markets for high-bandwidth wireless communication services. The ITU's IMT-2000 initiative may one day converge, but today, many 3G proposals are still under consideration including:

- cdma2000 (an upgrade to cdmaOne)
- *Universal mobile telecom system* (UMTS)
- *Wideband-CDMA* (W-CDMA)
- *Universal wireless communications* (UWC-136)

UWC-136 is based on TDMA as are Europe's GSM, Japan's *personal digital cellular* (PDC), and the *digital advanced mobile phone system* (D-AMPS) used in the U.S.

Existing 2G service providers have already applied for licenses to operate 3G networks around the globe. Although it's unclear what 3G technologies will be adopted, the most 2.5G upgrades are GPRS and *high-speed, circuit-switched data* (HSCSD), an upgrade being considered by some GSM network operators. Beyond that, EDGE modulation extensions are planned, which will enable service providers to offer even higher performance, enabling true 3G-like services.

The ITU currently embraces various proposed schemes to attain the IMT-2000 3G vision. From TDMA-based 2G providers of GSM and *North American dual-mode cellular* (NADC) services, interim upgrades will come in the form of GPRS, HSCSD, and IS-136+, and will eventually converge at EDGE for the next throughput upgrade (to 384 Kbps) before 3G.

What Is Special About EDGE?

EDGE is a new modulation scheme that is more bandwidth efficient than the GMSK modulation scheme used in the GSM standard. It provides a promising migration strategy for HSCSD and GPRS. The technology defines a new physical layer: 8-phase shift keying (8-PSK) modulation, instead of GMSK. 8-PSK enables each pulse to carry 3 bits of information versus the GMSK 1-bit-per-pulse rate. Therefore, EDGE has the potential to increase the data rate of existing GSM systems by a factor of three.

EDGE retains other existing GSM parameters, including a frame length, eight timeslots per frame, and a 270.833-kHz symbol rate. The GSM 200-kHz channel spacing is also maintained in EDGE, enabling the use of existing spectrum bands. This fact is likely to encourage deployment of EDGE technology on a global scale.

UMTS

Universal Mobile Telecommunications System (UMTS) is a part of the International Telecommunications Union's "IMT-2000" vision of a global family of 3G mobile communications systems. UMTS will play a key role in creating the future mass market for high-quality wireless multimedia communications that will approach 2 billion users worldwide by the year 2010.

UMTS is a modular concept that takes full regard of the trend of convergence of existing and future information networks, devices, and services, and the potential synergies that can be derived from such convergence. UMTS will move mobile communications forward from where we are today into 3G services, and will deliver speech, data, pictures, graphics, video communications, and other wideband information directly to people on the move. UMTS is one of the major new 3G mobile communications systems being developed within the framework that has been defined by the ITU and known as IMT-2000.

Over the past decade, UMTS has gained the support of many major telecommunications operators and manufacturers because it represents a unique opportunity to create a mass market for highly personalized and user-friendly mobile access to tomorrow's untethered society.

UMTS will build on and extend the capability of today's mobile technologies (like digital cellular) by providing increased capacity, data capability, and much greater range of services. The launch of UMTS services will see the evolution of a new, open communications universe, with players from many sectors coming together to deliver new communications services, characterized by mobility and advanced multimedia capabilities. The successful deployment of UMTS will require new technologies, new partnerships, and the addressing of many commercial and regulatory issues.

UMTS will enable tomorrow's wireless knowledge worker, delivering high-value broadband information, commerce and entertainment services users via fixed, wireless, and satellite networks. UMTS will speed the convergence between voice, data, and multimedia to deliver new services and create fresh revenue-generating opportunities. UMTS will deliver low-cost, high-capacity mobile communications offering data rates up to 2Mbit/sec with global roaming and other advanced capabilities.

The next decade will see the emergence of 3G networks to fully realize mobile multimedia services. Enabling anytime, anyplace connectivity to the Internet is just one of the opportunities for 3G networks. The major market opportunity builds on mobile networking to provide the following:

- Group messaging
- Location-based services (GPS)
- Personalized information
- Infotainment

Many new 3G services will not be Internet-based—they will be truly unique mobility services. Data will increasingly dominate the traffic flows. Pent-up latent demand for mobile data services will jump start 3G networks. By 2005, more data than voice will flow over mobile networks. This is an amazing statistic considering that mobile cellular networks today are almost exclusively voice.

Mobile Internet—A Way of Life

The mobile Internet is about to enter our daily lives in a big way. It will change the way we keep in touch with our friends and family, the way we do business, the way we shop, the way we access entertainment, and the way we conduct our personal finances.

The Internet is already a part of daily life for most of us, giving us access to a vast range of information and online services from our desktop computers. As a way of conducting business, it is also of growing importance to the global economy. Unlike today's fixed Internet, the mobile Internet will give us access to these services and applications wherever we are, whenever it suits us, from personal mobile devices.

By 2004, there will be as many as 600 million users of mobile Internet services. This means that more people will use mobile Internet than fixed Internet. The market is already taking off. The chart shown in Figure 10-6 represents the 3G subscribers projected for the future.

Eight to 10 billion SMS messages are sent worldwide every month. In Japan, there are more than 10 million users of the iMode service—which is comparable with basic WAP service—and each week another 150,000 new iMode users are added.

In a few years, many of us will wonder how we managed without the mobile Internet; it will become an invaluable part of our everyday lives. It will give us more opportunities to keep in touch with friends, family, and colleagues; empower us to make fast, yet well-informed business decisions; give us instant access to information and services and enable us to purchase the things we need or desire—all in a handy, pocket-sized device. We can expect to see the following types of services from 3G services:

- Customized Infotainment
- Multimedia Messaging Service
- Mobile Intranet/Extranet Access
- Mobile Internet Access
- Location-based Services
- Rich Voice

Rich Voice A 3G service that is real-time and two-way. It provides advanced voice capabilities (such as *Voice over IP* [VoIP], voice-activated net access, and Web-initiated voice calls), while still offering traditional mobile voice features (such as operator services, directory assistance, and roam-

Figure 10-6
Worldwide 3G
subscribers in millions

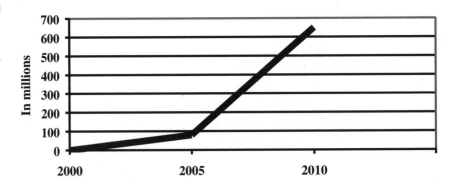

ing). As the service matures, it will include mobile videophone and multimedia communications.

At present, the mobile network value chain is centered on the network operator who captures more than 90 percent of market revenues, dominated by income from voice-based services. It is widely recognized, however, that advancing technology, growth of Internet services, and new end-user demands are challenging this traditional value chain.

The new, fast changing value chain will have new players and entities, and many network operators are already adopting new business strategies to broaden their role and to defend their competitive position. The multimedia service provider will be one of the key players in the multimedia value chain. Revenues will increasingly be diverted to other market players than the traditional.

The success of 3G will not just come from the mere combination of two existing successful phenomena-mobility and the Internet. The real success of 3G will result from the creation of new service capabilities that genuinely fulfill a market need. Meeting market demand is not just a question of technological capability and service functionality. Creating and meeting market demand requires services and devices to be priced at acceptable levels. This requires economies of scale to be present. The ability to benefit from economies of scale is one of the strongest market drivers for 3G services. *Universal Terrestrial Radio Access* (UTRA) now includes both the Direct Sequence and Time Code components of IMT-2000 and so embeds both the FDD access mode previously known as W-CDMA as well as the TDD modes previously known as TD-CDMA and TD-SCDMA. UTRA is now applicable to the major markets of Europe, China, South Korea, and Japan. UMTS promises significant economies of scale.

The need to protect existing investments in different 2G technologies has shifted the drive toward a single global standard. When you add the significant event of the emergence of the Internet, the additional capabilities of 3G became more focused on the provision of high data rates to deliver multimedia services. The emergence of the Internet as a mass-market content resource had justified the need for such high data rate capabilities and has since shifted the emphasis to packet-switched, IP-based core networks. There is general acceptance within the industry that 3G core networks will eventually be all-IP based.

The solution was the introduction of the IMT-2000 family of systems concept for 3G. One consequence of that solution is that a single global standard does not exist yet. However, the UMTS Forum believes that progress of technology, operational deployments, and market requirements will continue toward convergence. Another consequence — important when considering market perspectives—is that 3G now means different things in different parts of the world.

In Europe, 3G refers to the UMTS technology members of the IMT-2000 family, derived from GSM and deployed on new spectrum. A strong focus within the UMTS community is on international roaming capabilities and the potential benefits of the economies of scale that result from a common standard deployed across many nations. The same UMTS technology members will be used in South Korea, China, Japan, and most of the Asian region. In the U.S., 3G refers to derivatives of existing 2G technologies, deployed largely on occupied spectrum. 3G in the U.S. focus more on high data rates; international roaming capabilities are not a significant concern. The U.S. has lagged behind other world regions in the deployment of 2G digital cellular. Industry opinion is that it will continue to lag behind in the deployment of 3G. In Japan and South Korea, 3G means an opportunity to join the worldwide opportunity.

In 2G technologies, GSM currently has 65 percent of the world market shown in Figure 10-7. Japan has decided that its PDC 2G technology will not be evolved to 3G but will be replaced by the UMTS/IMT-2000 technologies. The TDMA and GSM communities are working on harmonization procedures in the approach to 3G. The 15 percent of the world market currently using cdmaOne technology, mainly located in the U.S. and South Korea, has a transition path to the IMT-MC member of the IMT-2000 family but is limited to existing spectrum.

With UMTS services, providers worldwide will be using multi-band rather than multi-mode handsets—a much more attractive proposition for terminal manufacturers.

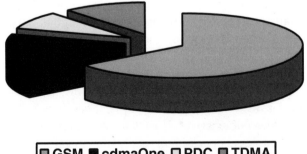

Figure 10-7
GSM user population worldwide

☐ GSM ■ cdmaOne ☐ PDC ☐ TDMA

Applications of the Wireless Internet

Using the mobile (wireless) Internet as the model for the growth curves seen, the following are some of the applications for moving to a wireless and a 3G environment:

- **Cutting the umbilical cord**   The first wave is making our familiar on-line services mobile—"cutting the cord" of the Internet. An example is using a laptop computer together with a mobile phone to send and receive e-mail or surf the Net. Users can access Internet-based services at the airport, on a train, or in the park.

- **Pocket WWW**   The second wave, which has already begun, brings Internet services to pocket mobile devices. Applications are specially adapted to work on mobile devices with small screens, for example, using WAP. Although this wave brings full, convenient mobility, it is still largely based around traditional Internet services, such as on-line banking, e-mail and web access.

- **Real mobility in the Internet**   In the third wave, the full potential of the mobile Internet is realized. Services, applications, and content are centered on the mobility, location, and situation of the user—they become "situation-centric."

This intelligence can be used to create highly valuable, personalized services. Mobile devices will become indispensable tools that enhance our daily lives. Services will be relevant, useful, and timely. Figure 10-8 is a forecast

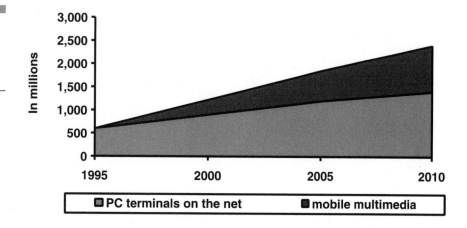

Figure 10-8

Growth figures of mobile multimedia devices on the Internet

of the growth and changes that will occur on the wireless Internet devices over the next decade.

Visions of Wireless

The convergence of high-speed wireless data communications, "always-on" mobile computing platforms, and instant access to the Internet is driving the biggest shift in mobile computing since the advent of the computer itself. The day is coming when the mobile professional will be linked wirelessly to the power of the Internet, no matter where on the globe he may find himself. He/she will need instant access to critical corporate, personal and public data applications, such as e-mail, e-commerce, stock trading, weather, airline reservations, hotels, car rental agencies, and sports information. The GSM/GPRS community becomes the gatekeeper to the needs of the true mobile professional.

There will be a consolidation in devices as manufacturers seek to meet customer needs while controlling costs. Smart phones and *Personal Digital Assistants* (PDAs) will become talkative PDAs. 3G Laptops and 3G Web Tablets, while keeping their separate identities, will both be handheld Internet devices, hitting the high and low end of 3G users. Like all of the 3G devices, the 3G Web Tablet will have a product lifecycle of its own through 2010. As innovative technology and customer demand cause the accumulation of all new device capabilities, the 3G Multimedia Device or 3G Personal Companion will become the sought after all-in-one mobile Internet tool for the middle of the decade. Others have predicted that by 2010, there will be a single wireless gadget that will meet all needs.

- **Smart Phones/WAP Phones** These early devices provide content and Web browsing. They use standard and new operating systems and protocols (like Pocket PC and WAP), and will soon synchronize with other devices (like desktops and mobile phones). As WAP becomes popular and takes advantage of the high data rates and "always on" capability that GPRS will provide, these devices will naturally evolve into some of the first 3G devices at the even higher 3G data rates. The Smartphone will evolve to a Talkative PDA.

- **Personal Digital Assistant (Talkative PDA)** Although there is room for improved coverage and quality, today you can purchase a PDA that also has mobile voice communications (for example, radio modems for GSM, OmniSky). Besides their calendars, address books, and other organizing features, PDAs are thin and lightweight. Many have color screens and are quickly gaining computer strength due to low power chip designs, screen miniaturization, and evolving operating systems. As they grow in computing capability while maintaining their hand-held form factor, they will continue to distinguish themselves from 3G laptops as less expensive, less powerful solutions. Examples are numerous with Palm, Casio, HP, and others leading the pack. The Kyocera phone and PDA combination is a new twist in the movement toward the PDA.

- **3G Laptop (Handheld Internet)** Laptops today have modems and *Personal Computer Memory Card International Association* (PCMCIA) cards that enable wireless communications. They continue to get smaller, lighter, and with more powerful computing. With the bandwidth offered with 3G, these powerful, portable computers will thrive with the custom graphics, two-way video, and large file transfers for the future.

- **3G Web Tablet (Handheld Internet)** Appearing in 2000 "Wireless Web Tablets," these devices offer portable Internet access by plugging into power and access at home, and gaining limited mobility via a short wireless connection. As low cost, lightweight, thin Internet appliances the size of magazines, these devices offer e-mail, robust Internet access and, Web browsing. Eventually, they will gain both full mobile access and synchronization with other devices via more powerful 3G spectrum.

- **3G Multimedia Device (Personal Companion)** Today's slow connections based on low bandwidth cause "jerky" video images. Compression techniques cannot overcome the need for speed and capacity. 3G will answer this problem in the mobile world. There are many visions of the ultimate 3G devices with some saying it will evolve from phones, others from computers. Because there will be different 3G services addressing specific user needs, all of the previous devices will develop from both worlds. However, there will be a need for an all-powerful device that does quality VoIP, full Internet access, and two-way video.

In order to understand the role of next generation wireless services in the broader technology landscape, it is important to understand the current state of the Internet industry and other enabling technologies that shape its development. The Internet is transitioning from an inexpensive medium for advertising, marketing, and customer support to a common platform for transactions and business applications. At the same time, technological and commercial developments are melding together information, communications, commerce, and entertainment into one large, consolidated industry. Part of the reason for this evolution is because more consumers are accessing the Internet using multiple devices and over multiple communications networks. They are also changing their behavior and consumption patterns. In addition, tools and facilities are available that improve the consumer Internet access experience.

Wireless access to the Internet is going to drive the overall development of the Internet for several reasons:

- Wireless enables service providers and Internet businesses to increase their mobile culture and total service consumption.

- The mobility and immediacy offered by wireless enables Internet content delivery and commerce to be location non-specific.

- The person-specific nature of wireless enables companies to develop customer profiles that enable them to narrowcast and distribute better value-added information to customers.

- Location-based facilities and services provide another tier of customer knowledge that enables Internet businesses to deliver context-specific services that also improve customer value.

In short, wireless is an opportunity for Internet businesses to learn more about their customers, understand their customers' consumption patterns, strengthen their customer relationships, and provide more personalized services. This is a critical component of Internet business strategies and what wireless operators/service providers bring to the table in a full Internet solution.

The most important lifestyle users who use wireless access are as follows:

- **Business professional** This includes the high value mobile users (such as busy decision-makers). The services used will include intranet access, messaging, and scheduling systems.

- **Product managers** These are users with specific occupational requirements for high volumes of information while in a mobile mode. Requirements include remote and mobile access to corporate and external information.

- **Youth** Often, early adopters of technology will be inclined to use instant messaging, games, and entertainment-oriented services.

- **Parents** In many countries, both parents work and share the household responsibilities.

- **Senior citizens** 3G will enable more reliable support electronically and reduce the requirement for labor intensive support services such as medical monitoring, location based medical service, family, caretakers and caregivers, and social workers.

Positioning the Mobile Industry

In the light of the new business chain, the issue is to consider whether to simply provide a wireless IP pipe to a service offering hosted elsewhere on the Internet, or to go for an interoperable end-to-end solution. The wireless IP pipe business using tunneling protocols will become a commodity operation, where cost, coverage, and data rate are the only competitive dimensions. By carefully developing and pre-selecting useful Internet-based mobility services with competitive tariffs, the user will be encouraged to buy into UMTS services.

At issue is the location of the subscriber profile records, which reflect the personalized service choices of the end-user: message filtering options, choice of mobile information and type of mobile device, correlated with name, billing address, mobile phone number, and e-mail address. This store of data will permit additional returns through selectively targeted mobile e-commerce and advertising. UMTS Operators have three distinct separate or combined possibilities:

- *They charge the subscriber on a metered call basis.* This makes it possible to get a revenue stream via a small additional charge for mobile Internet services. It produces a return on investment well before mobile commerce and advertising become feasible. Added to the traffic revenue, the operator can capture or selectively share this revenue with value chain partners on its own terms.

- *The Operator provides IP-packet transport (such as GPRS-based).* All is necessary to integrate Internet services with Intelligent Networking, voice, data, and fax services. Volume discounts will become a possibility beyond billing for just time.

- *The Operator will know the subscriber's location using emerging cellular positioning technologies.* Positioning adds end-user value through information customization, such as details of the nearest restaurant or automatic conversion of e-mail to speech for a driver of a moving car. In the future, location information will enormously increase the revenues.

Wireless Internet will become one of the media channels for content providers, and wireless network operators will join in on the service values.

The WAP-Portal offerings of GSM operators and iMode offerings of NTT DoCoMo are examples of a new strategy. The mobile Portals are unique because they are a solution in which operators and service providers can manage content and integrate with communication and transactions.

The arrival of WAP and iMode are generally seen as the first steps of the convergence between the Mobile Telecommunications, the Internet, and content industries.

Key Technologies

Some of the critical technologies essential for the successful introduction of UMTS include the following.

UTRA

The ETSI decision in January 1998 on the radio access technique for UMTS combined two technologies — W-CDMA for paired spectrum bands and TD-CDMA for unpaired bands — into one common standard. This powerful approach promises an optimum solution for all the different operating environments and service needs. The transmission rate capability of UTRA will provide at least 144 Kbps for full mobility applications in all environments, 384 Kbps for limited mobility applications in the macro and micro cellular environments, and 2.048 Mbps for low mobility applications particularly in the micro and picocell environments. The 2.048-Mbps rate may also be available for short range or packet applications in the macro cellular environment, depending on deployment strategies, radio network planning, and spectrum availability.

Multi-Mode Second Generation/ UMTS Terminals

UMTS terminals will exist in a world of multiple standards, and this will enable operators to offer maximum capacity and coverage to their user base by combining UTRA with second and other third generation standards. Therefore, operators will need terminals that are able to interwork with legacy infrastructures, such as GSM/DCS1800 and DECT, as well as other second-generation worldwide standards such as those based on the U.S.

AMPS standard, because they will initially have more complete coverage than UMTS. Many UMTS terminals will therefore be multi-band and multi-mode so that they can work with different standards, old and new. Achieving such terminals at a cost, which is comparable with contemporary single mode second-generation terminals, will become possible because of technological advances in semiconductor integration, radio architectures, and software radio.

Satellite Systems

At initial service launch in 2002, the satellite component of UMTS will be able to provide a global coverage capability to a range of user terminals. These satellite systems are planned using the S-band *Mobile Satellite Service* (MSS) frequency allocations identified for satellite IMT2000 and will provide services compatible with the terrestrial UMTS systems.

USIM Cards/Smart Cards

A major step forward, which GSM introduced, was the *Subscriber Identity Module* (SIM) or smart card. It introduced the possibility of high security and a degree of user customization to the mobile terminal. SIM requirements, security algorithms, and card and silicon IC technology will continue to evolve up to and during the period of UMTS deployment.

The smart card industry will offer cards with greater memory capacity, faster CPU performance, contact less operation, and greater capability for encryption. These advances will enable the *UMTS Subscriber Identity Module* (USIM) to add to the UMTS service package by providing portable high security data storage and transmission for users, as well as configuration software for the operation of any UMTS terminal, images, signatures, personal files, fingerprints or other biometric data that could be stored, down- or up-loaded to or from the card.

Contact-less cards will permit much easier use than with today's cards, for example, enabling the smart card to be used for financial transactions and management such as electronic commerce or electronic ticketing without having to be removed from a wallet or phone. It is expected that all fixed and mobile networks will adopt the same or compatible lower layer standards for their subscriber identity cards to enable USIM roaming on all networks and universal user access to all services. Electronic commerce and banking using smart cards will soon become widespread and users will expect and be able to use the same cards on any terminal over any network.

New memory technologies can be expected to increase card memory sizes making larger programs and more data storage feasible. Several applications and service providers could be accommodated on one card. In theory, the user could decide which applications/services he wants on the card, much as he does for his computer's hard disk. This is the challenge and opportunity for service industries that evolving smart card technology presents.

Internet Protocol (IP) Compatibility

UMTS is a modular concept that takes full regard of the trend towards convergence of fixed and mobile networks and services, enabling a huge number of applications to be developed. As an example, a laptop with an integrated UMTS communications module becomes a general-purpose communications and computing device for broadband Internet access, voice, video telephony, and conferencing for either mobile or residential use.

The number of IP networks and applications are growing fast. Most obvious is the Internet, but private IP networks (intranets) show similar or even higher rates of growth and usage. UMTS will become the most flexible broadband access technology, because it allows for mobile, office, and residential use in a wide range of public and non-public networks. UMTS can support both IP and non-IP traffic in a variety of modes including packet, circuit switched and virtual circuit.

UMTS will be able to benefit from parallel work by the *Internet Engineering Task Force* (IETF) while they further extend their basic set of IP standards for mobile communication. New developments like IP version 6 enables parameters such as *quality of service* (QoS), bit rate and *bit error rates* (BER), vital for mobile operation, to be set by the operator or service provider. Developments on new domain name structures are also taking place. These new structures will increase the usability and flexibility of the system, providing unique addressing for each user, independent of terminal, application, or location.

UMTS has the support of many major telecommunications operators and manufacturers because it represents a unique opportunity to create a mass market for highly personalized and user-friendly mobile access to the Information Society. UMTS seeks to build on and extend the capability of today's mobile, cordless, and satellite technologies by providing increased capacity, data capability, and a far greater range of services using an innovative radio access scheme and an enhanced, evolving core network.

Spectrum for UMTS

WRC 2000 identified the frequency bands 1885 to 2025 MHz and 2110 to 2200 MHz for future IMT-2000 systems, with the bands 1980 to 2010 MHz and 2170 to 2200 MHz intended for the satellite part of these future systems. *Code Division Multiple Access* (CDMA) is characterized by high capacity and small cell radius, employing spread-spectrum technology and a special coding scheme.

The capabilities of *cdmaOne* evolution have already been defined in standards. IS-95B provides ISDN rates up to 64 Kbps. The next phase of cdmaOne is a standard known as 1XRTT, which enables 144-Kbps packet data in a mobile environment. Other features available include a two-fold increase in both standby and talk time on the handset. All of these capabilities will be available in an existing cdmaOne 1.25-MHz channel. The next phase of cdmaOne evolution will incorporate the capabilities of 1XRTT, support all channel sizes (5 MHz, 10 MHz, and so on), provide circuit and packet data rates up to 2 Mbps, incorporate advanced multimedia capabilities, and include a framework for advanced 3G voice services and vocoders, including voice-over-packet and circuit data. Many of the steps have already been started and put in place.

A number of flavors of CDMA are shown in Table 10-1.

The cdma2000 Family of Standards

The cdma2000 family of standards includes core air interface, minimum performance, and service standards. The cdma2000 air interface standards specify a spread spectrum radio interface that uses CDMA technology to meet the requirements for 3G wireless communication systems. In addition, the family includes a standard that specifies analog operation to support dual-mode mobile stations and base stations.

Purpose

The technical requirements contained in cdma2000 form a compatibility standard for CDMA systems. They ensure that a mobile station can obtain service in a system manufactured in accordance with the cdma2000 standards. The requirements do not address the quality or reliability of that

Table 10-1

Variations of CDMA

CDMA Type	Description
Composite CDMA/TDMA	Wireless technology that uses both CDMA and TDMA. For large-cell licensed band and small-cell unlicensed band applications. Uses CDMA between cells and TDMA within cells.
CDMA	In addition to the original Qualcomm-invented N-CDMA (originally just "CDMA") also known in the US as IS-95. Latest variations are B-CDMA, W-CDMA, and composite CDMA/TDMA. CDMA is characterized by high capacity and small cell radius, employing spread-spectrum technology and a special coding scheme. B-CDMA is the basis for 3G UMTS.
cdmaOne	First Generation Narrowband CDMA (IS-95).
cdma2000	The new second-generation CDMA MoU specification for inclusion in UMTS.

service, nor do they cover equipment performance or measurement procedures. Compatibility, as used in connection with cdma2000, is understood to mean: Any cdma2000 mobile station is able to place and receive calls in cdma2000 or IS-95 systems. Conversely, any cdma2000 system is able to place and receive calls for cdma2000 and IS-95 mobile stations. In a subscriber's home system, all call placements are automatic. Similarly, call placement is automatic when a mobile station is roaming. To ensure compatibility, radio system parameters and call processing procedures are specified. The sequence of call processing steps that the mobile stations and base stations execute to establish calls is specified, along with the digital control messages and, for dual-mode systems, the analog signals that are exchanged between the two stations.

The base station is subject to different compatibility requirements than the mobile station. Radiated power levels, both desired and undesired, are fully specified for mobile stations, in order to control the RF interference that one mobile station can cause another. Base stations are fixed in location and their interference is controlled by proper layout and operation of the system in which the station operates. Detailed call processing procedures are specified for mobile stations to ensure a uniform response to all base stations. Base station procedures, which do not affect the mobile stations' operation, are left to the designers of the overall land system. This approach to writing the compatibility specification is intended to provide the land system designer with sufficient flexibility to respond to local service needs and to account for local topography and propagation conditions. cdma2000 includes provisions for future service additions and expansion of system capabilities. The release of the cdma2000 family of standards supports Spreading Rate 1 and Spreading Rate 3 operation.

Wireless Applications

Using Wireless

What can we do with all these different standards and technologies? Why is it so important to have so many different choices? Why should I care? What are the applications that will demand the bandwidth that we have been discussing throughout this book?

All of the questions that continue to arise regarding broadband wireless communications are very valid. Make no mistake about this; the applications that will stress the bandwidth of today do not yet exist. However, as in all cases of technology advancements, we face the chicken and the egg dilemma. Many folks argue that few (if any) applications will use the high-speed communications. Others argue that if the bandwidth would be available, the applications would quickly follow. So we end up in the proverbial argument about supply and demand. Build it and they will use it!

How we approach this argument is what sets the systems and the programmers apart. Standards bodies, manufacturers, and carriers all have an obligation to stay ahead of the demand curve. They look out 3 to 5 years and determine what might be available if the protocols and standards are set. They also look at what the demand will be in the 5-year and beyond window. If they wait until the demand rises, they will be behind the curve and be under constant pressure. Conversely, if they jump too quickly, they will be ahead of the curve and the demand. Either way, the price of being ill-prepared is too great to withstand. Many network providers in the past have suffered due to lack of service or facilities and paid the price in money or their very existence. How many CLECs were too fast on the trigger in offering services over an infrastructure that was not there? The fallout is only beginning to appear. Many CLECs offering high-speed access over wireline services (CATV, xDSL, and so on) have gotten into financial difficulty. The difficulty these carriers experienced was not of having too little too late, instead, they had too much too soon. The RBOCs deliberately waited until these providers blew their budgets and their capital, and then swooped in to clean up after the emerging providers.

Nearly all of the trade press agencies have had a field day on the progress of the emerging applications and the subsequent development of broadband wireless services. These magazine and news agencies all profess that the acceptance rate for wireless is lackluster. Although some merit rests in these statements, the one issue being overlooked is the delay in getting to a standard. If in fact the standards lag, then the products will also lag behind. Manufacturers do not want to develop products that are non-standard (although they are quick to add a proprietary spin on their individual products) due to the risk of being displaced or losing their market share. Now we see another issue in as much as the products are not rolling out as fast as they should, so the pricing model is still too high for mass markets. As pioneers start to experiment with the technologies, protocols, and products, they do so with the understanding that the prices are higher in the experimental stages than in mass production. We have been in commercial trials for many of the protocols, technologies, and individual products for a couple of years, and more will follow. We have also seen the escalation of the various rollouts of technological improvements. These situations combined create sheer havoc in the industry, paralyzing the vendors who do not know which product or technology to select. *Fear, uncertainty, and doubt* (the FUD factor) reign.

There Is a Bug in My Soup

As of this writing, several issues continue to plague the industry rollout of *third-generation* (3G) wireless solutions. In May 2001, British Telecommunications PLC delayed the launch of what was planned as the world's first commercial network using the 3G mobile telecommunications technology. The plan was to gain the edge over competition by at least three months. The delay is attributed to a "software bug" with handsets from their suppliers. The result of this glitch cost BT Telecommunications a minimum of 3 to 6 months.

The bug, a code problem, in the integrated software caused the NEC-made handsets to crash. The problem occurs between the base station and the handsets, which couldn't be fixed in time. Specifically, when a mobile phone user moves between base stations, the connection is automatically cut off.

In April 2001, the industry discovered that Nokia phones were not compatible with the 3G services that were rolling out across the world.

French telecommunications equipment maker Alcatel does not expect 3G UMTS mobile phones to hit the market before 2004. Alcatel, which recently launched its first *general packet radio service* (GPRS) mobile phone to speed up online services on GSM networks and plug the gap before UMTS, blamed the delay on the time and investment needed to develop the technology. According to Alcatel, the rollout of 3G will take 3 to 5 years (2004 to 2006), whereas last October they were saying 2 to 3 years.

Not everything seems to be as smooth in the overall evolution as it appears. However, problems and glitches will occur in any innovation and growth curve. 3G developments are slowed, but they will amass support and the engine will roll in the future. Together with the expected increase in data transmission rates that will be around 40 times that of today's GSM and TDMA by 2003 or 2004, the 3G initiatives and the 3G networks will be the building blocks of a true wireless information society. The distinctions between mobile phones, fixed telephones, portable computers, and personal digital assistants will disappear in the future. 3G devices will be multi-tasking, handling up to six different simultaneous calls, whether data, video, or voice. These applications and products are the tip of the market's iceberg.

In April 2001, Japan's largest cellular telecommunications carrier, NTT DoCoMo Inc., also pushed back the launch of its 3G services to work out some technical problems. NTT DoCoMo said that software glitches were behind its decision to limit its initial offering of a super-fast new cell phone technology. NTT DoCoMo had repeatedly promised to start 3G wireless service by late May. That would have made the Tokyo company the world's first to offer such commercial service, a supercharged wireless that was supposed to eventually turn cell phones into miniature computers for surfing the Net, downloading music, and watching movie reviews.

UMTS terminals will take over in 2007, not 2005, as originally anticipated. This means that the GPRS life cycle will be longer than originally anticipated. GPRS was a short-term solution to UMTS when it first rolled out. However, now it appears that the GPRS sets will be used to satisfy longer lead times and slightly slower data transfers at up to 170 Kbps.

Wireless Intenet Starts It Off

The wireless Internet is well on its way to establishing a new market for wireless handsets, PDAs, laptops, and other wireless clients, as wireless operators begin to deploy high-speed data services in mobile networks worldwide (that is, GPRS, WCDMA, EDGE). The wireless handset shipments were not that poor in 2000. The wireless handset market did not grow as planned, because there were few new services and features offered. The wireless handset basically remained a voice tool. However, that will begin to change soon. Data-enabled handsets utilizing WAP and i-mode™ will account for 80 percent of the handsets shipped in 2006, up from 30 percent in 2000. Handsets utilizing 2.5G or 3G technology will rise to over half of all handsets shipped in 2006, with WCDMA having the largest share of the group. Bluetooth will also play a large role in wireless connectivity, with over 90 percent of all handsets shipped, incorporating Bluetooth by 2006.

Including Bluetooth will allow wireless clients to access the WAN, LAN, and PAN.

Developers are planning on a multi-faceted attack with the devices to support wireless applications for the future. A survey of manufacturers indicated that these developers were planning two major thrusts (handsets and PDAs, followed by pagers). This survey is shown in Figure 11-1, representing the results of the manufacturer's targets. The numbers will add up to more than 100 percent because the manufacturers are making multiple devices.

Applications and Features

Initial users of DoCoMo's service are not charged for the handsets but pay for the transmission fees of about 1.8 times the current charge with the higher fee being about 80 cents or $1.25 for a 3-minute song. Three handset models are available, including the much-touted videophone that will show the caller on the other end of the line on the tiny cell phone screen. The videophone is shown in Figure 11-2.

The model will also be able to show other types of video. At the onset, users will be able to look at video that is part of DoCoMo's image-distribution service, offering more than 100 programs such as news and short film clips for a video gadget called Eggy, shown in Figure 11-3.

Figure 11-1
Device manufacturers' plan for wireless access service

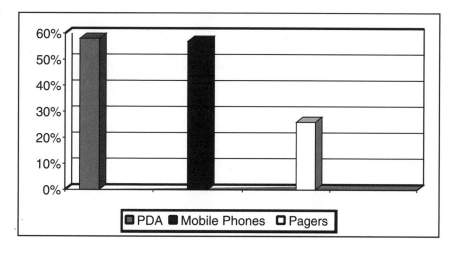

Figure 11-2
Video Phone
(Source: Nokia)

Figure 11-3
Eggy
(Source: DoCoMo)

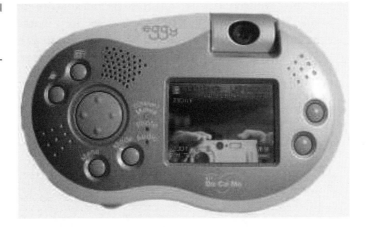

Another 3G model is a faster version of the current NTT DoCoMo Internet-linking i-mode phones with nearly 23 million users in Japan. The third model will be a card-type device strictly for data transmission. NTT DoCoMo plans to offer 3G in both Europe and the U.S. in the future. It plans to offer its i-mode in the U.S. as early as 2002 through its partner AT&T Wireless. Plans also account for some parts of Europe in late 2001 or early 2002. A representative graph of the wireless Internet users reveals that Japan leads the industry followed by Korea, Europe, and the U.S. Currently, the world's wireless Internet users are distributed, as shown in Figure 11-4.

What this means is that the U.S. and Europe are clearly behind Japan when it comes to wireless data transfer. This is not a race, but an explanation of why the conversation tends to lean toward the NTT DoCoMo model and the manufacturing comes primarily from Japanese vendors. The rest of the word is quickly becoming interested in high-speed wireless Internet access using this model and the i-mode service.

Wireless support for most applications is becoming paramount. The number of wandering workers and road warriors consistently grows. In surveys of the Fortune 500 companies, research houses have found that 77 percent of these companies provide remote access to traveling staff and 74 percent support telecommuters. Wireless access is targeted to be a top priority for upwards of 85 percent of these organizations as the demand for instant communications from anywhere rises. Interestingly, 44 percent of the organizations indicated that handheld devices (palm tops, notebooks, and sub-notebooks) are used for corporate network access. The survey also went on to determine that 78 percent of the Fortune 500 companies actually pay for the handheld devices. However, security is an issue that continually crops up when dealing with wireless access. Figure 11-5 is a graph of the number of respondents who felt that wireless access (and broadband access to be more specific) would be a top priority over the next few years.

Figure 11-4
Number of wireless
Internet users

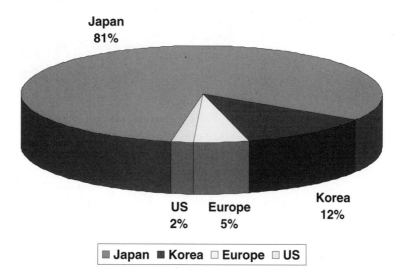

Japan 81%

US 2%   Europe 5%   Korea 12%

■ Japan ■ Korea □ Europe □ US

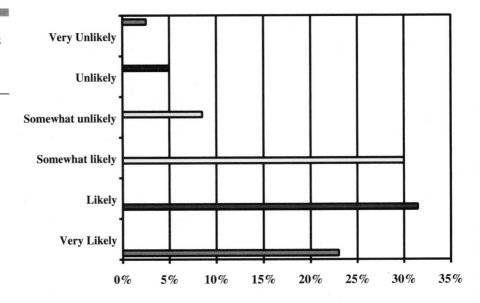

Figure 11-5
Opinions of wireless
importance
(Source: Network
world)

Figure 11-6
Plans to provide
wireless Web access

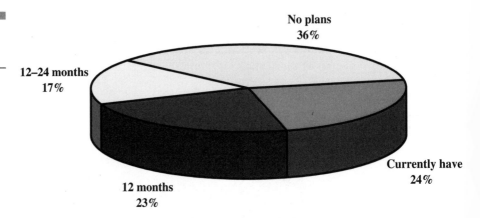

Another area the surveys were interested in determining were the plans for corporate users in the top 500 companies to implement wireless Web access. The survey looked at a time frame of 2 years. The results of that inquiry are shown in Figure 11-6.

TV as an Application

Samsung Electronics is now marketing a cdma2000 1x mobile handset with the capability to receive motion pictures in color. The *video on demand* (VOD) phone was completed in 2000. The handset supports VOD, *audio on demand* (AOD) and other advanced functions slated for IMT-2000. It is expected to spark intense competition among content providers. The Color VOD Phone comes with a 2.04-inch TFT-LCD that can reproduce clear motion picture images in 200,000 different color shades. The handset can send and receive text data wirelessly at up to 144 Kbps as well as support VOD and AOD. Thus, mobile access is possible to various content services with color moving pictures such as music, videos, Internet broadcasts, animation, news reports, and so on.

The handset is based on cdma2000 1x and comes with a MPEG4 motion picture decoder and stereo player built in. It can display 12 lines of text at once. A separate memory is available to download motion picture clips for playback. The energy-efficient nature of the system and optimized software combine to greatly extend battery life to provide ample time for viewing motion pictures in color.

The latest Samsung cellular phone presents a significant change of the mobile technology by switching from the black-and-white display and transmission of data only to a color display and VOD phone. By the introduction of the VOD phone in the market, the world will enjoy the IMT-2000 service, futuristic technology for the human being. Samsung will market the new VOD phone alongside an STN-LCD model that can reproduce 256 different colors.

The STN-LCD model works from the cdma2000 1x platform and can send and receive text, data, and color images at up to 144 Kbps. The Samsung phone can be used to access games and various other mobile services and is expected to quickly replace handsets with black-and-white screens. The video set is shown in Figure 11-7.

Interestingly, the use of a TV receiver is not one of the initial applications that people wanted. In reality, the application of a videoconferencing system was more appropriate for a business service. However, things change and the culture of the end user changes. Now that the TV receiver can be integrated into the wireless handset, we can expect that two-way multimedia with high-end graphics using an MPEG-4 scenario will become the new demand.

Figure 11-7
The color TV receiver
in a handset
(Source: Samsung)

What About Dick Tracy?

Although the handheld streaming video reception for TV may have some appeal, others may choose a wireless communicator that is less sophisticated. Perhaps these users may want a high-speed data receiver and a cell phone combination that take no added space on the desk or in the pocket. Enter the wristwatch phone! These devices have some ergonomically designed features that let you send and receive from your arm, without having to stop and find desk or table space. The wristwatch phone is shown in Figure 11-8.

What if we could include a video camera in the watch? Today, watches are available that double as a camera, and we have watches that double as a phone. So why not combine the features together using Bluetooth and WAP, for example, to use the service of the threesome? Suppose that you need to take a picture for insurance purposes. An accident occurs, so you take a picture immediately of the two vehicles involved and immediately transmit

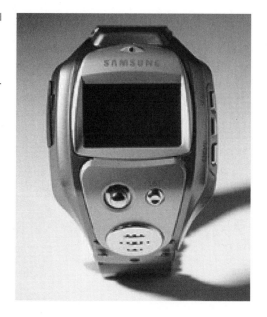

Figure 11-8
The wristwatch
phone
(Source: Samsung)

that picture to the police to file a record of the incident. Simultaneously, you send the information to the claims adjuster at your insurance company so that they have a record of the event and a picture of the damages. This may speed the process of filing a claim and prevent the mishandling of estimates to fix the damages. The combination is not 100 percent there yet, but it is a matter of time until the pieces are combined together into a useful tool.

Web Through the Sky

Before we go too far out, the use of broadband wireless communications (greater than 56 Kbps) can also give us more access from the sky, via satellite. Many of the ISPs are looking at how they might offer us higher speed access than what we are accustomed to on the landline networks. This is not a handheld application right now but may likely become on in the future with the use of the LEO technologies, as we discussed with Teledesic and Iridium LLC. Earthlink (a national ISP who acquired Mindspring and Netcom) has arranged to use the services of DirecPC to offer its customers high-speed Internet access. This service will offer up to 400 Kbps downlink and 144 Kbps uplink capabilities to customers using a 2- by 3-foot dish, as

Figure 11-9
Satellite access to
the Internet

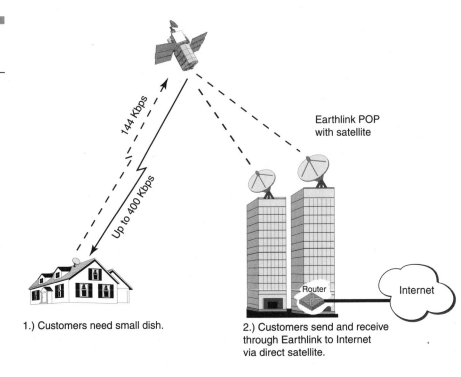

1.) Customers need small dish.

2.) Customers send and receive
through Earthlink to Internet
via direct satellite.

shown in Figure 11-9. This differs from the original DirecPC (also called the
DirecDuo in earlier versions) where the original offering from the digital
broadcasters was a one-way downlink at up to 400 Kbps and a dialup connection to the ISP over the landline. This version includes two-way communications across the satellite link. The facility is a shared resource
similar to the cable TV companies who offer Ethernet services to the Internet, but speeds are dependent on the number of simultaneous users on the
network. The WAP protocol or the iMode protocols will be integrated into
the wireless Web environment.

The direct connection to the satellite can still be handled through a satellite provider by using the connection similar to EarthLink, but using a dial
up connection as described with the *Digital Broadcast Service* (DBS), as
shown in Figure 11-10. In this case, the application is again given access to
the Internet. However, the user will dial on a reverse link into a satellite-
based ISP. The dial-up connection is slower speed communications at
speeds of 33.6 to 56 Kbps. At the ISP, a connection is made to the Internet
server. The difference here is that on the forward link (from the ISP back to
the user) the transmission is sent via satellite. The ISP redirects the pack-

Figure 11-10
The dial-up
connection to an ISP
is received via
satellite.

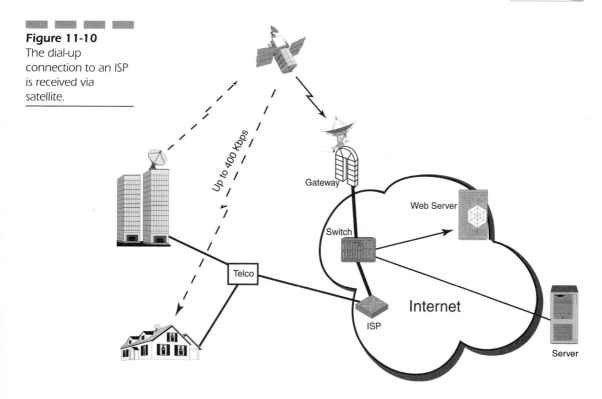

Figure 11-10
The dial-up
connection to an ISP
is received via
satellite.

ets to a connection over the satellite link back to the user at a much higher
speed (usually up to 400 Kbps).

Earlier in Chapter 10, we discussed the potential pitfalls of TCP/IP over
a satellite link. The problems are not insurmountable, but the consideration
of latency is ever present. We have known since the 1960s that satellite
technology is an excellent way to transmit our high-speed data communications. Because the satellites are in such a high orbit (the GEO), the need
for multiple hops is reduced, possibly to a single hop through the airwaves.
The existing satellite orbits in use can literally provide access to anywhere
in the world due to the footprint and coverage areas available at this height.
However, whereas TCP/IP was originally designed to work over any
medium, the response times for TCP over a satellite link can severely
degrade the throughput. Newer gateways have been introduced to minimize the effect of the satellite delays and enhance performance. The gateways are used as intermediaries between the end user connection and the
satellite link. Functionally, the TCP data is captured by the gateway. The

gateway transmits the data across the satellite channel that has been optimized for TCP and then a new connection is established on the receiving side by a gateway on the other end. The connection is therefore comprised of three parts, as shown in Figure 11-11:

- A TCP connection between the client and a remote gateway
- A TCP connection across the satellite between two gateways (client and server)
- A connection between the server gateway and the server (receiver)

This is designed to be fully transparent to the end user and should reduce the latency on the overall circuit. This improves the chances of satellite communications being used for broadband wireless connectivity.

Figure 11-11
Satellite gateways improve TCP performance.

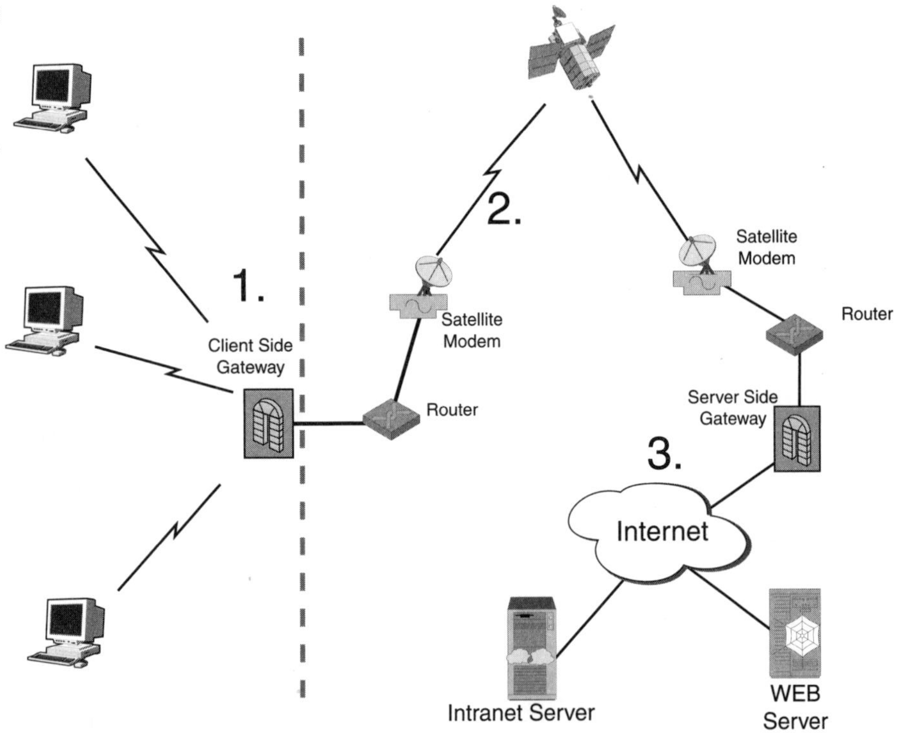

Through the Air with Non-LOS

Several other applications can be linked together with broadband wireless communications. Not all sites can guarantee that *line of sight* (LOS) is available. Therefore, a non-LOS system can satisfy some of the applications needed. Take, for example, the needs in a community to link multiple sites together because of communities of interest. The library, town hall, and police and fire departments need interconnectivity to each other but may not have all the access necessary. Herein lies the problem; not all sites are on the same access links and LOS may not be available. The cost of leased wireline circuits may also be prohibitive.

Enter the broadband wireless solution using an LMDS or MMDS connection. Fixed wireless communications can be very effective as a solution in the township. Customers use a small dish receiver and a modem at each site and connect via the MMDS services on a broadcast capability up to 2 Mbps. Using LMDS architecture, similar services can support speeds today up to 155 Mbps and 622 Mbps. Using a free space optics connection, the OC-3 or OC-12 in free space can satisfy most current needs for voice, data, Internet access, and multimedia communications at a price that is very competitive.

Obviously, many organizations and municipalities are catching on to the idea of using the ISM band (2.4 GHz unlicensed) frequencies to provide the connections they need. A recent survey of the MMDS users using broadband wireless shows that the use of MMDS is on the rise, as shown in Figure 11-12.

Figure 11-12
Broadband wireless
users on MMDS

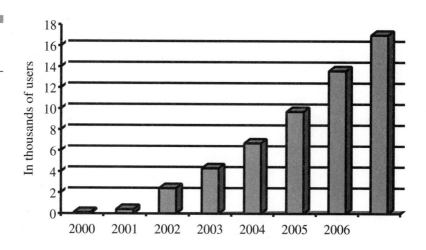

Medical Prescriptions

Let's look at a few other applications that really have some merit. First shown in Figure 11-13 is a graphic that shows an operating room scenario. The medical staff is in the operating room preparing a patient for surgery. Before surgery is performed, the doctors need a biopsy of some tissue that was sent to the pathology lab. The pathologist performs the necessary tests and immediately sends the results through a wireless connection (through an access point in the lab and one in the operating room) to the doctor for analysis. At the same time, the results may be input wireless to a database engine for inclusion into the patient's records. Now the doctor may need additional information from the server. So, a query is sent to the database server for an analysis of

- Prognosis or diagnosis
- The procedures normally performed
- Recommendations or concerns from other surgeons

Figure 11-13
Medical applications for broadband wireless communications

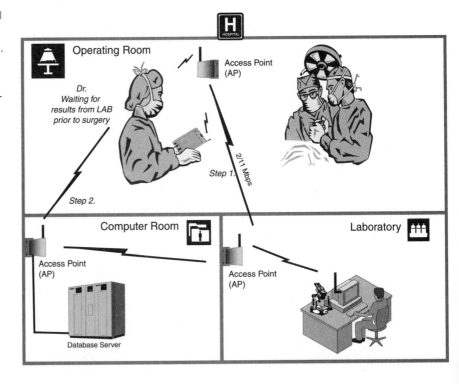

- *Medical resonance imaging* (MRI) images
- Vital statistics needed

These analyses are not uncommon; the difference is that they are being performed immediately so that the doctor can provide the necessary emergency treatment without the wait normally experienced by having someone run around to gather the information.

During this same cycle, the doctor may also look up the patient's medial records and past procedures to make sure the procedure about to be performed is not going to jeopardize the patient. An example may be that there is no drug interaction that may be a risk, or that the patient had a similar operation earlier. The benefit is timely information where you need it and when you need it. Perhaps this sounds like it is a low-bandwidth demand rather than a broadband need. In this instance, we use high-speed communications at 11 Mbps or 54 Mbps (depending on the wireless LAN used) to gather the information quickly. However, graphical and MRI files may also consume bandwidth and cause inordinate delays. This is a lifeline service where the demand is for fast and reliable data communications.

Although this case example is using the service in a hospital environment, cases will occur where the medical staff is not supportive of the wireless communications. They feel that the wireless transmissions will interfere with the other patients' lifeline services such as pacemakers, radio-controlled wheelchairs, and the like. These incidents are few and far between and more operations are embracing the technology.

Another scenario includes the interference within a medical building not allowing the data to pass through walls and floors, especially where the radiology lab, for example, uses lead lining to prevent radiation from escaping the lab. This is why the access points are set in the individual rooms for ad hoc communications. The access point is wireless, the handheld (palm or PDA) is wireless, but the backbone network may be wired. This is a mix of the broadband cable and wireless networks merging together.

A midwest medical institution currently uses these wireless palm devices for doctors to update medical charts while making their rounds through the hospital. All too often, the specialists visit their patients and keep very accurate records on the individual. These are later transcribed and kept on file. The problem occurs when the transcriptionist doesn't understand some of the information or mistypes the information. Still another problem occurs when the records are filed. Misfiled records account for many disappointments. When the records cannot be found, the patient may have to go through tests and documentation again.

Now with the use of a wireless *personal digital assistant* (PDA), the information is entered into the terminal device during a physical exam. Medical history, problems, diagnoses, and treatment plans are all directly input at the time and are transmitted directly to the LAN server via a wireless access point. Any instructions are immediately sent to the nurses' station or other doctors in the hospital/ clinic. Entries can be made once, as shown in Figure 11-14, during the doctor's rounds, and misspelling and misinterpretation is minimized. The PDA can actually be used to pull down a script with all the necessary information in a format that is easy for the doctor to fill in with either the graffiti or stylus entry. If the doctor needs MRI or X-ray information during the rounds, these will be immediately accessible in the future over the broadband wireless communications to the PDA.

Dental Floss and PDAs

Another scenario is the use of the technologies already discussed in previous chapters of this book. Bluetooth is being built in to cell phones, PDAs, palm devices, and future systems. As a communications protocol, it is really

Figure 11-14
The PDA is used to input during rounds.

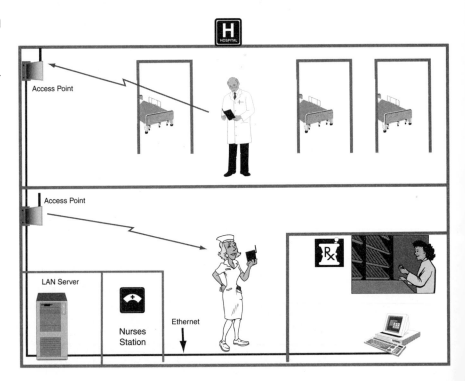

a replacement to the wires that are traditionally used. These combined devices (PDA and Bluetooth protocols) will enable the use of the palm in a school environment. Many universities are now issuing Palm or Visor devices to the students as a staple item. The Palm shown in Figure 11-15 and the visor shown in Figure 11-16 are commonly found in schools and in road warriors' potpourri of tools.

Figure 11-15
The Palm
(Source: Palm)

Figure 11-16
The Visor
(Source:
DoCoMo/Handspring)

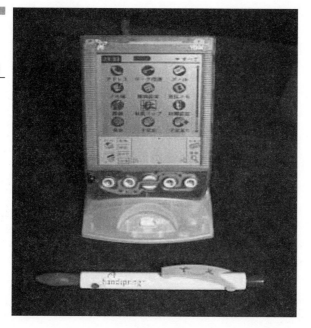

Figure 11-17
The handheld
Palmtop PC
(Source: IBM)

An alternative to the Palm or Visor is to use a handheld PC (sub-notebook), as shown in Figure 11-17. In this case, the device is used like any PC with a keyboard entry. The wireless LAN card can operate at 2, 11, or 54 Mbps using the 802.11 standards.

SOHO Long, It's Been Good to NOYO

As the broadband wireless networks of the future continue to mature, the use of combinations of protocols and devices becomes very possible. Bluetooth and WiFi (WLAN) will work together in the *small office or home office* (SOHO). Between the home office and the vehicle, users can select from myriad devices where content is created, used, transmitted, or stored. Using a portfolio of wireless access methods, the *personal area network* (PAN) can be constructed to work between the desktop and the LAN across a wireless network.

Two basic building blocks are used to create this PAN for the home office-networking scenario: 802.11b, also called WLAN, and Bluetooth. Both operate in the ISM band (operating at 2.4 GHz). Using the WLAN, we can communicate up to 11 Mbps, whereas Bluetooth can support speeds of up to 1 Mbps for data, plus three 64-Kbps voice channels. This combination in the PAN prevents interference and enables us to use both within the same areas.

Other devices are designed to work in the small office or home office and be compatible with the 802.11b specification operating at 11 Mbps. These devices can enable a connection from the office to the Internet across a DSL link, with a four- or eight-port Ethernet Hub operating in one or many boxes. An example of the wireless SOHO access point is the SOHOware Net Blaster II that operates through walls, ceilings, and floors at distances of 150 to 500 feet, as shown in Figure 11-18.

Students and Profs Come Together

The wireless devices of the future will enable you to issue commands to the PDA from a cell phone, or have the PDA speak with a laptop device. The wireless world is coming fast. However, beyond just issuing commands, the

Figure 11-18
SOHOware's
Netblaster II
(Source: SOHOware)

wireless communications will handle the high-speed bandwidth requirements and multimedia requirements for future applications. Meanwhile, back to those schools that issue the PDA (palms) to all students. The devices will be used for interconnectivity between the student and the professor via e-mail, and students can instantly request an appointment with a professor when they need guidance or have issues to resolve. The professor can synchronize the e-mail application and receive mail wherever he or she is. Moreover, a response can immediately be sent to the student and be synchronized to the professor's calendar. As an addition to this, the students can collaborate with each other or download important Web site or research information to the handheld palm device for later use. The download of Web site information is shown in Figure 11-19.

However, downloads are not just limited to students. Road warriors may also need software or Web downloads while they are mobile. This can happen while they are on the road from a vehicle, from a hotel room, or from their homes. Figure 11-20 is a representation of software downloads from a vehicle as an example of how this may happen. Note that in a broadband

Figure 11-19
Students will be able to download from the Web or school database.

Download of Web site info
@ 128–384 Kbps

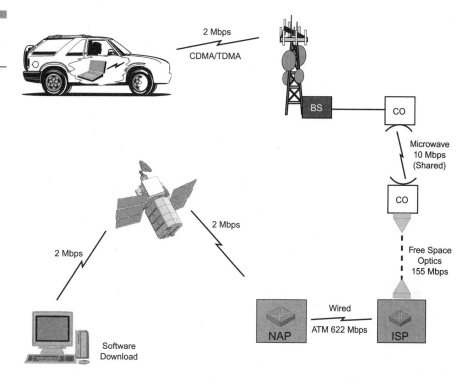

Figure 11-20
Downloads for road warriors

wireless environment, the 2 Mbps throughput is what we are looking for today, but higher speeds will be a plus.

Shopping Panorama

Department stores can benefit from this integration of broadband wireless communications. A department store may want to increase sales during a lull period. As an example, a tent sale, as shown in Figure 11-21, can be held in the parking lot. Instead of running LAN wires out to the cash registers in the lot, a wireless communications device can be used. That is easy and proven technology. Symbol Technologies and others have been offering wireless registers and scanners for years.

Now, however, we want to scan the barcode of an item, enter that information into the active wireless register, and also deduct the items from inventory in the back room. Simultaneously, the order entry function may

Figure 11-21
Store access to
Intranet and
Extranets

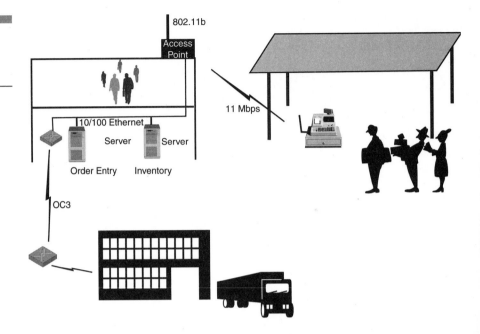

generate a requisition to replace a depleted item in the store's inventory, which in turn generates a purchase order to replace the items. This immediately is transmitted to the manufacturer's warehouse (Extranet) and causes the items to be picked and packed for immediate shipping. This order entry function, full cash management, inventory management, and replacement function all occur from the cash registers outside the store in the lot, while at the same time, the registers inside the store are active. The demands for this instant access will grow as more just-in-time deliveries are used in the industry, holding the inventory to a bare minimum.

In Your Face

Earlier in this chapter, a cell phone with a TV on it and a video conferencing capability were seen. That same video conferencing capability can be added to the handheld PDA (sub-notebook, palm, or visor) and a 384-Kbps speed (or higher) videoconference can occur on demand. The palm top is shown with a lipstick camera attached, plus a wireless modem to facilitate the access to EDGE or 3G networks for real-time video. The video can be

Figure 11-22
*Real-time video
conferencing on a
palm makes the
service instantly
available.*

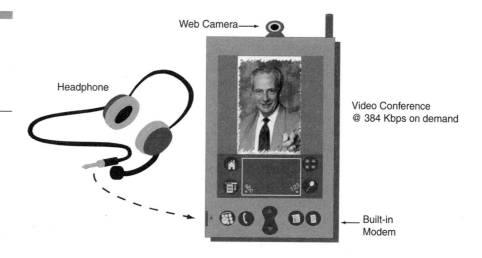

Web Camera →

Headphone

Video Conference
@ 384 Kbps on demand

Built-in
Modem

using an MPEG compression but can also produce 30 real-time frames of video per second, giving us a smooth video play, as shown in Figure 11-22. The use of this service is fairly easy and reasonably priced or it will not be used.

One vendor has already discussed real-time videoconferencing from a cell phone or PDA at normal rates for the future ($.10-per-minute USD). This remains to be seen but offers some exciting possibilities in areas like field service, testing, interviewing, and depositions that can be taken any-time from anywhere. This PDA concept makes the overall wireless network easy to access. Moreover, it enables us to drop the device in a briefcase, purse, or pocket. Some of us use the belt clips to carry our PDA around so the availability is always on.

Instead of using a palmtop device, a multimedia capability can also be used from a notebook or sub-notebook computer where the user may have a Web site up in one window, a videoconference on another, and a complex document in still another window. The collaborative sharing of documents with a videoconferencing capability, as shown in Figure 11-23, is an exam-ple of the higher bandwidth demands for wireless communications. Here the end user may be a SOHO or telecommuter working from home or a road warrior calling in from a hotel room. Unfortunately, few hotels offer any high-speed services to the Internet or a dial-up service at broadband speeds; therefore, the wireless access must be portable enough to take with you.

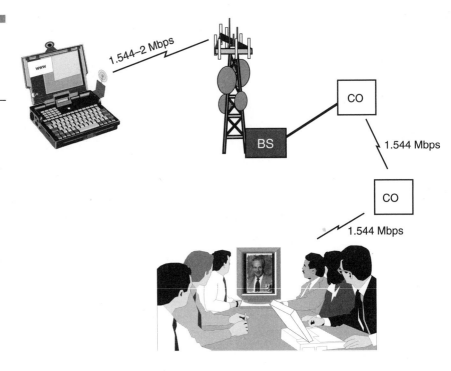

Figure 11-23
Multimedia on a
notebook with
wireless connections
at 1.5 to 2 Mbps

Ad Hoc Meetings with InfraRed

We cannot ignore that in a dynamic workplace the need will always be present to set up an ad hoc working committee or team. Regardless of whether it is product design, project management, marketing, or sales, these teams need the ability to share information. Many times these teams are brought together in remote sites so that their work can be handled confidentially, or where they can get the most accomplished without day-to-day interruptions. All too often we try to book a service (such as a hotel or retreat area) where these people can work and share their information. In many cases, high-speed interconnectivity is just not available, limiting the places where the meetings can be held. In other cases, a hodge-podge of interconnections and wiring schemes may be used, deterring from the actual work at hand. As a result, wireless intercommunications in an ad hoc environment, as shown in Figure 11-24, work extremely well. The results are inexpensive connections, equal connectivity for the entire group, and a better work product. This is effective with long-term committees, but it will also work for impromptu meetings and document sharing.

File sharing
File transfer

Figure 11-24
Sharing in an ad hoc
meeting with RF or
InfraRed

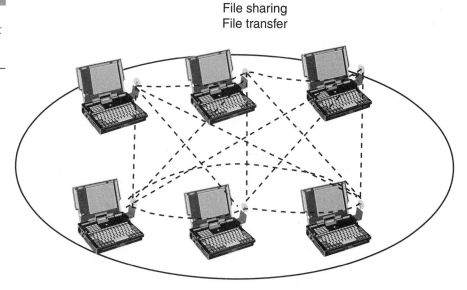

File sharing
File transfer

Finally the Set Changes

Just so we do not slant this discussion too much to the right and discuss the PDAs only, such as the palm and visor (and others), we should look also at the changes that are occurring on the telephone (cellular) sets. In reality, the sets are morphing as more powerful telephones but are also taking on the PDA-like functionality shown in Figure 11-25. Some phones like the PCS ones from Kyocera include a PDA in the telephone, making the set

Figure 11-25
The phone and PDA
combination offer
high-speed Internet
and telephony
(Source: Kyocera).

more accessible and allowing a Web-enabled capability. The PCS architecture using a CDMA access is said to allow access at speeds of up to 144 Kbps today but will evolve to be fully cdma2000-compliant at 2 Mbps. The telephone (Smartphone) opens to reveal a PDA tool kit that enables voice-activated dialing and Web access; however, it is powered by the Palm OS, incorporating the functions of the palm, phone, and pager all in one. This is a shift in one direction. Others will likely follow with devices that match closely in a CDMA and TDMA environment.

Phoneless in Chicago

Motorola, on the other hand, chose to move away from the telephone set and introduce its Accompli, which is a mobile phone that doesn't look like a phone. It is also a PIM and offers trimode GSM and GPRS protocols, phone functionality, Internet access, e-mail, and *short message service* (SMS) with a full QWERTY keyboard and a 256-color screen, as shown in Figure 11-26. Poised to capitalize on the rollout of GSM/GPRS networks in North America, the Accompli offers business users a first look at the new world of handheld communications that will be available with the introduction of

Figure 11-26
The Accompli doesn't look like a phone (Source: Motorola).

global networks and high-speed data access to North America. It operates with the Wisdom OS and supports about 30 different applications. With an infrared data port, users can send and receive important documents by fax, print to compatible printers, or beam business cards to compatible devices.

Where Am I Going?

Another application that is becoming more exciting as we progress is the use of the next generation of wireless device for location services. Using a GPS-type service, we can map our current location and trace out our routes of where we want to go. Now when we use such a service we are looking for the application to be crisp and robust. It would do no good to have a slow speed data transfer rate when we need directions now. Therefore we can get almost real-time location capabilities using a GPRS or EDGE service at 170 to 384 Kbps. This also works in vehicle-tracking applications where a dispatcher may want to determine the location of a truck to pick up a load of product. By using the location services triangulated with a GPS service, the application becomes a reality.

Moreover, this location service could be a chipset built in to the vehicle for tracking in the event of theft or loss. The vehicle's actual location needs to be real time if we plan to catch up with it (or report it to the authorities). It would do no good if we paint the location on the screen on a slow-speed connection and it takes minutes to load the information down. By the time the authorities arrive, the vehicle will be long gone.

Alternatively, the checking of information such as airline and train schedules plays an equal role in the use of the device. The handheld device with access to a mapping and scheduling service, as shown in Figure 11-27, is an example of what Nokia is doing for this application. We can actually do an online booking and catch the next available train to wherever we want to go.

How Do I Get There from Here?

Inevitably, we all will want to get to a point where we can use the high-speed intercommunications devices and networks. In some cases, we have to be patient while the operators build out their networks and bring them to us. In other cases, we can begin by experimenting with in-house broadband

Figure 11-27
Location services on
a high-speed
interface
(Source: Nokia)

communications using a WLAN or HomeRF network. Still, we may look at using some of the emerging features in sporadic locations because ubiquity is still a few years off. But just how do we plan the migration strategy? We can take some steps to prepare for the broadband wireless networks in-house and out-house. These steps include involving management and users alike and planning the budgets.

Begin by setting a management strategy on what the migration will be like and what departments, applications, and users will be involved. The steps may include the following, as shown in Table 11-1. You can add or delete any items to this list appropriately within your own organization. The list is not exhaustive but sets the stage for developing the strategy.

At the end of a trial and rollout, the three questions we should be able to answer are as follows:

- What is it?
- What is it going to do for me?
- What is it going to cost?

No matter what other icing we put on the issues, the questions and answers always boil down to these three factors. If the questions cannot be answered, you better go back to the drawing board.

Table 11-1

Summary of Steps
for Planning a
Broadband Wire-
less Network

Step	What You Are Trying to Accomplish
Conduct a needs assessment.	1. What are the management's expectations with a wireless access plan? 2. What do users think they need? 3. What will be the impact of adding a lower speed communications network to a high-speed wired network? 4. What can be accomplished differently with wireless that could not be done before? 5. Any licensing issues? 6. Are there certain access methods that are allowed, and others not allowed? 7. Will VPNs be used? 8. How will you control access? 9. Is real estate required to house the transmitters?
Develop a preliminary plan.	1. Do the necessary research regarding the technologies, such as WLAN, HomeRF, and WIFI. 2. Understand the differences between RF and light-based systems. 3. Talk with vendors about where the products are in relation to standards. 4. Are there interoperability needs or issues? 5. Is the equipment standard(s) compliant? Are there any plans in the future that would change that?
Develop a preliminary budget.	1. Understand the approximate costs per user. 2. Look at the possible choices (high-end and low-end), speed, and access methods. 3. What can be used during the test bed and what can be kept later? 4. What is the training cost necessary to bring users up to speed? 5. What tools do you need that are different? (cost and human) 6. Are there new applications to be introduced? (cost) 7. Is the company going to issue the tools or is it a user responsibility to buy?
Set a trial period.	1. After determining the application, install the software on the PCs or other devices. 2. Order the components (hardware). 3. Set a trial period no more than reasonable (60 to 90 days if localized). 4. Set the departments to be involved. 5. Decide how to back up the applications, the components (handheld devices, cards, and so on), and the access methodologies. 6. Determine what makes the trial a success. What are the factors that will be improved? 7. Does the plan look for cost increases or decreases? 8. Is access the factor that will influence management or users?

(continued)

Table 11-1

Summary of Steps
for Planning a
Broadband Wire-
less Network
(continued)

Step	What You Are Trying to Accomplish
Run the test.	1. Establish a coordinator who is responsible for making sure the test is run according to the plan. 2. Set a procedure for reporting errors, problems, and other issues. 3. Make sure that the data is secure, and access is hack-proof. 4. Start rolling out the application on a department-by-department basis. 5. Introduce complications if possible to see if they can be easily resolved. 6. Test the vendor's capacity to respond to problems. 7. Check users for their opinions and defuse hot issues as soon as possible. 8. Document everything you do.
Plan the mass market training.	1. Train the IS personnel in the operation and programming requirements. 2. Train users on how to use the applications and services. Include any new logon procedures and any shim software for VPN training if selected. 3. Train the technical staff on how to fix hardware and software problems on the network.
Complete the test.	1. Know when to say it is completed, and compile the results of the test. 2. Document the planned results and the actual results. 3. Prepare a debriefing on what works and what does not. Be prepared to discuss why something did not work. 4. Assess personnel input, department management's input, customer relationship management improvements, and other factors. 5. Have any of the success factors changed?
Roll out.	1. If the test was a success, then plan to roll it out en mass. 2. Add new departments one at a time. 3. Follow up on all results with each new department added. 4. Document any changes. 5. Keep track of the budget. 6. Keep management informed. 7. Keep customers informed. 8. Finalize all training on a department-by-department basis. 9. Look for new applications that make business sense.
Maintain the network.	1. Don't stop. Keep checking everything. 2. Add new or updated equipment. 3. Upgrade applications as needed. 4. Test equipment on hand (if you maintain it yourself). 5. Follow up reports to management.

Final Comments

We have traveled down a very long and winding road. Several tributary routes occur along the way (technology differences) and obstacles (interoperability issues) occasionally pop-up. However, if we think back to where this all started, I wanted to be sure that you saw the various options available, the opportunities to increase productivity, and the ways of selecting a workable solution with which you could feel some comfort. Too many choices are available, each with their own positive and negative points. You, the user, must make your own decision, but now you may be better equipped to ask the right questions of the vendors.

There are no real right and wrong answers. You must make a choice; if it works for you, it is correct! If it does not work for you, then make a new one. Don't feel that you need to lock yourself into a path that has no escape routes. Pick the wireless broadband solutions as they become available to you. And by the way, try to have some fun while you use these services. Imagine using a handheld or a wrist-mounted wireless device that can be multiple productivity tools all rolled up into one. That is the excitement of the wireless world. Good luck!

ACRONYM LIST

2G	Second generation; generic name for second generation of digital mobile networks (such as GSM)
3G	Third generation
3G-CDMA	Third-generation wireless CDMA
3GPP	Third-generation Partnership Project
3GSM	Third-generation GSM
A5	GSM privacy algorithm
AC	Authentication Center
ACA	Automatic channel assignment
ACK	Acknowledgement
ACTS	Advanced Communication Technology Satellite
ADPCM	Adaptive Differential Pulse Code Modulation
AM (or ASK)	Amplitude Modulation (or Amplitude Shift Keying)
AMPS	Advanced Mobile Phone Service
AP	Access Point
API	Application Programming Interface
APN	Access Point Name
ASP	Application Service Providers
ATDMA	Asynchronous Time Division Multiplexing Access
AUC	Authentication Center
BARG	Billing and Accounting Rapporteur Group
BCS	Block Check Sequence
BER	Bit Error Rate
BG	Border gateway
BGP	Border Gateway Protocol
bps	Bits per second
BSC	Base Station Controller
BSS	Base Station Subsystems
BSSGP	BSS GPRS Protocol
BSSMAP	Base Station System Mobile Application Part
BST	Base Station Transceiver
BTS	Base Transceiver Station
CAGR	Compound Annual Growth Rate
CAMEL	Customized application of mobile enhanced logic
CAP	Competitive Access Provider
CBC-IF	Cell Broadcast Center Interface
CBS	Cell broadcast short message service
CC	Control channel

CCA	Clear Channel Assessment
CDMA	Code Division Multiple Access
CDPD	Cellular Digital Packet Data
CEIR	Central Equipment Identity Register
CEPT	Conference of European Posts and Telecommunications
CMRS	Commercial Radio Service
CO	Central Office
COCC	Constellation Operations Control Center
COMP 128	A3/A8 authentication and cipher key generation algorithm
CRC	Cyclic Redundancy Check
CRTC	Canadian Radio and Television Commission
CS	Circuit-switched; cell station or cell site
CSD	Circuit-switched data
CSG	Communications Strategy Group
CTIA	Cellular Telecommunications Industry Association
DAMPS	Digital AMPS
DBMS	Database Management System
DCA	Dynamic channel assignment
DCCH	Digital Control Channel
DCF	Distributed Coordination Function
DCS	Data Coding Scheme; Digital Cellular System
DFA	Designated Filing Area
DHCP	Dynamic Host Configuration Protocol
DIFS	Distributed Inter-Frame Spacing
DL	Downlink
DM	DataTAC Messaging
DNS	Domain Name System
DS	Direct sequence
DSI	Digital Speech Interpolation technique
DSSS	Direct Sequence Spread Spectrum
DTD	Document Type Definition, definition of a language built on XML or SGML
EBRC	EMC and Bio-effects Review Committee
EC	Executive Committee
ECSA	Exchange Carriers Standards Association
EDACS	Enhanced Digital Access Communications System
EDGE	Enhanced data rates for GSM evolution
EIA	Electronic Industries Association
EIR	Equipment Inventory Register
EMC	Electro-Magnetic Compatibility

EME	Electromagnetic Energy
ETDMA	Extended Time Division Multiple Access
ETSI	European Telecommunications Standards Institute
FCC	Federal Communications Commission
FDD	Frequency Division Duplex
FDMA	Frequency Division Multiple Access
FF	Fraud Forum
FH	Frequency Hopping
FHMA	Frequency Hopping Multiple Access
FM (or FSK)	Frequency Modulation (or Frequency Shift Keying)
FMS	Fraud Management System
FSS	File Specification Subgroup
FWRA	Fixed Wireless Radio Access
G_b	Interface between a SGSN and a BSS
G_c	Interface between a GGSN and a HLR
GC	Group Chairmen (group)
GCF	Global Certification Forum
G_d	Interface between a SMS-GMSC and a SGSN, and between a SMS-IWMSC and a SGSN
GEO	Geosynchronous orbit
G_f	Interface between a SGSN and an EIR
GGRF	GSM Global Roaming Forum
GGSN	Gateway GPRS Support Node
GHz	Gigahertz
G_i	Reference point between GPRS and an external packet data network
GIWU	GSM interworking unit
GMSC	Gateway mobile services switching center
GMSK	Gaussian Minimum Shift Keying
G_n	Interface between two GSNs within the same PLMN
G_p	Interface between two GSNs in different PLMNs
GPDS	General Packet Data Services
GPRS	General Packet Radio Systems
GPS	Global Positioning System
G_r	Interface between an SGSN and an HLR
G_s	Interface between a SGSN and an MSC/VLR
GSM	Global System for Mobile Communication
GSMA	GSM Association
GSN	GPRS Support Node (xGSN)
GTP	GPRS Tunneling Protocol
GTR	Group Trailer, which indicates the end of packet group

GUTS	General UDP Transport Service
GW	Gateway
HDLC	High-Level Data Link Control
HDML	Handheld Device Markup Language
HF	High frequency
HLR	Home Location Register
HSCSD	High-Speed Circuit Switched Data software upgrade for cellular networks that gives each subscriber 56K of data
HTML	HyperText Markup Language, document definition language of WWW
HTTP	HyperText Transfer Protocol, transfer protocol of WWW
HUR	High Usage Report
Hz	Hertz
iDEN	Integrated Digital Enhanced Network
IE	Information Element
IETF	Internet Engineering Task Force
IIDB	Interworking Issues Database
IIF	Interworking and Interoperability Function
IMEI	International Mobile Equipment Identity
IMSI	International Mobile Subscriber Identity
IMT2000	International Mobile Telephony 2000
IMTS	Improved Mobile Telephone Service
IOT	Inter-Operator Tariff
IP	Internet Protocol
IPR	Intellectual Property Rights
IR	Infrared
IrDa	Infrared port
IREG	International Roaming Expert Group
IRTF	Interstandard Roaming Task Force
IS	Information Services
ISG	IMT 2000 Steering Group
ISM	Industrial Scientific Medical
ISP	Internet Service Provider
IST	Information Services Technology
IWF	Interworking Function
IWU	Interworking Unit
kHz	Kilohertz
L2TP	Layer two Tunneling Protocol
LAN	Local area network
LAP-D	Link Access Protocol-Data Channel
LAPi	Link Access Protocol iDEN

LEO	Low earth orbit
LLC	Logical Link Control
LOS	Line of sight
LPC	Linear Predictive Coding
LSB	Least significant bits
MAC	Medium Access Control
MAHO	Mobile-assisted handoff
MAP	Mobile Application Part
MDBS	Mobile Data Base Station
MDG	Mobile Data Gateway
MD-IS	Mobile Data-Intermediate System
MDLP	Mobile Data Link Protocol
MEO	Mid-earth orbit
MF-TDMA	Multi-Frequency Time Division Multiple Access
MGL	Maximum Group Length
MHz	Megahertz
MIS	Management Information Service or System Support
MM	Mobility management
MMI	Man Machine Interface
MMS	Multimedia Messaging Service
MPL	Maximum Packet Lifetime (constant)
MPS	Maximum Packet Size
MRP	Market representation partner (within 3GPP)
MS	Mobile Station (ME + SIM)
MSA	Metropolitan service areas
MSC	Mobile-services Switching Center (for PDC); Mobile Switching Center
MSISDN	Mobile Subscriber ISDN (Telephone number or address of device)
MSS	Maximum Segment Size
MTP	Message Transfer Part
MTSO	Mobile telephone switching office
NADC	North American Dual-Mode Cellular
N-AMPS	Narrowband Advanced Mobile Phone Service
NAM	Number Assignment Modules
NAS	Network access server
NIF	Network Interface Facility
NMS	Network Management System
NOC	Network Operations Center
NOCC	Network Operations Control Center
OA&M	Operations, administration, and management

OR	Optimized routing
OSS	Operations Support System
PACCH	Packet Associated/Acknowledgement Control Channel
PAGCH	Packet Access Grant Channel
PAN	Personal Area Networks
PBCCH	Packet Broadcast Control Channel
PBCH	Packet Broadcast Channel
PBS	Personal base stations
PCCCH	Packet Common Control Channel
PCCH	Packet Control Channel
PCIA	Personal Communications Industry Association
PCM	Pulse Coded Modulation
PCN	Personal Communications Network
PCS	Personal Communication Services
PCU	Packet control unit
PDA	Personal digital assistant
PDC	Personal Digital Cellular
PDCH	Packet Data Channel
PDLP	Packet Data Link Protocol
PDN	Packet data network
PDP	Packet Data Protocol
PDTCH	Packet Data Traffic Channel
PHS	Personal Handy Phone System
PHY	Physical
PLCP	Physical Layer Convergence Protocol
PLL	Physical Link sublayer
PLMN	Public Land Mobile Network
PM or PSK	Phase Modulation
PMD	Physical Medium Dependent
PN	Pseudo-random Noise
PNCH	Packet Notification Channel
POP	Point of presence, Population unit
POTS	Plain Old Telephone Service
PPCH	Packet Paging Channel
PPP	Point-to-Point protocol
PRACH	Packet Random Access Channel
PRD	Permanent Reference Document
PSK	Phase Shift Keying
PSPDN	Packet Switched Public Data Network
PSTN	Public-Switched Telephone Network
PTCH	Packet Traffic Channel

PTM	Point-to-multipoint
PTP	Point-to-point
PVC	Permanent virtual circuit
QAM	Quadrature and Phase Modulation
QoS	Quality of service
QPSK	Quadrature Phase Shift Keying
R&TTE	Radio and Telecommunications Terminals Executive
RADIUS	Remote Authentication Dial-In User Service
RAS	Remote Access Server
RF	Radio Frequency
RFCL	Radio Frequency Convergence Layer
RFL	RF sublayer
RITL	Radio in the Loop Concept
RLC	Radio Link Control
RLP	Radio Link Protocol
RTT	Round-Trip Time
SACCH	Slow Associated Control Channel
SAR	Segmentation and Reassembly
SAS	Security Accreditation Scheme
SATIG	Satellite Interest Group
SATK (or STK)	SIM (Application) Toolkit
SCAG	Smart Card Application Group
SCCP	Signal Connection Control Part
SCR	Standard Context Routing
SEA	Spokesman Election Algorithm
SERG	Services Experts Rapporteur Group
SG	Security Group
SGML	Standardized General Markup Language, a general markup language that can be adapted to different applications with DTDs like HTML
SGSN	Serving GPRS Support Node
SIG	Special Interest Group
SIM	Subscriber Identity Module
SLA	Service-level agreement
SMG	Special Mobile Group (of ETSI)
SMPP	Short Message Peer to Peer Protocol
SMR	Specialized Mobile Radio
SMS	Short Message Service
SMSC	Short Message Service Center
SMSCB	Short Message Service Cell Broadcast
SNDCP	Subnetwork Dependent Convergence Protocol

SNR	Signal to noise ratio
SONET	Synchronous Optical Network
SPT	Server Processing Time
SS7	Signaling System 7
SSAR	Simplified Segmentation and Reassembly
SSL	Secure Sockets Layer, a security protocol now known as TLS
TACS	Total Access Communications Services; Total Access Control System
TADIG	Transferred Account Data Interchange Group
TAP3	Transferred Account Procedure 3
TBF	Temporary Block Flow
TCAP	Transaction Capability Application Part
TCH	Traffic Channel
TCP	Transmission Control Protocol
TCP/IP	Transmission Control Protocol/Internet Protocol
TDD	Time Division Duplex
TDMA	Time Division Multiple Access also known as D-AMPS
TE	Terminal equipment
TIA	Telecommunications Industry Association
TIA/EIA	Telecommunications Industry Association/Electronic Industry Association
TID	Tunnel Identifier
TLLI	Temporary Logical Link Identifier
TLS	Transport Layer Security, security protocol formerly known as SSL
TS	Time slot
TSAP	Transport Service Access Point
TSC	Trunking System Controller
TTML	Tagged Text Markup Language
TTR	Transmission Trailer, indicates the end of transmission
UDCP	USSD Dialog Control Protocol
UDH	User-Data Header (see GSM 03.40)
UDHL	User-Data Header Length
UDL	User-Data Length
UDP	User Datagram Protocol
UHF	Ultra High Frequency
UL	Uplink
U_m	Interface between the MS and the GPRS fixed network part
UMTS	Universal Mobile Telecommunications System

UP	Unified protocol
USF	Uplink state flag
USSD	Unstructured Supplementary Service Data
USSDC	Unstructured Supplementary Service Data Center
UTRA	Universal Terrestrial Radio Access
UWCC	Universal Wireless Communications Consortium
VAS	Value-added services
VHF	Very High Frequency
VLR	Visiting Location Register
VoIP	Voice over Internet Protocol
VPLMN	Visitor Public Land Mobile Network
VPN	Virtual private network
VSAT	Very Small Aperture Terminal
VSELP	Vector Sum Excited Linear Predictors
WAE	Wireless Application Environment
WAN	Wide area network
WAP	Wireless Application Protocol
WARC	World Administrative Radio Conference
WASP	Wireless Application Service Provider
WCAP	Wireless Competitive Access Provider
W-CDMA	Wideband Code Division Multiple Access
WCS	Wireless Communications System
WDP	Wireless Datagram Protocol
WI	Wireless Internet
WISP	Wireless Internet Service Provider
WIT	Wireless Intelligent Terminals
WLAN	Wireless LAN
WLL	Wireless Local Loop
WORM-ARQ	WORM-Auto Repeat Request
WRC 2000	World Radio Conference 2000
WSP	Wireless Session Protocol
WTA	Wireless Telephony Application, application programming interface for controlling telephony features of the device
WTLS	Wireless Transport Layer Security, the security protocol of WAP
WTP	Wireless Transaction Protocol
WWAN	Wireless wide area network
WWW	World Wide Web
XML	Extendable Markup Language, a subset of SGML

GLOSSARY

3G (Third Generation) Wireless The next step in the development of wireless communications. The first generation was analog and the second was digital (CDMA, TDMA, and GSM). Third-generation systems are expected to provide broadband, high-speed data applications, both fixed and mobile.

Access Point A device that transports data between a wireless network and a wired network (infrastructure).

Acknowledgment The transmission of a short packet from the receiving device to the sending device to indicate that the data sent has been received error-free.

A/D Analog-to-Digital Converter (ADC) Converter that uniquely represents all analog input values within a specified total input range by a limited number of digital output codes.

AGC (Automatic Gain Control) A system that holds the gain and, accordingly, the output of a receiver substantially constant in spite of input-signal amplitude fluctuations.

AIN (Advanced Intelligent Networks) Refers to networks that route calls based on database information that can affect the inbound and outbound flow of the call.

Air Time Actual time spent talking on the wireless telephone. Most carriers bill customers based on how many minutes of air time they use each month. The more minutes of time spent talking on the phone, the higher the bill.

Alphanumeric A message or other type of readout containing both letters (alphas) and numbers (numerics). Regarding wireless, "alphanumeric memory dial" is a special type of dial-from-memory option that displays both the name of the individual and that individual's phone number on the wireless phone handset. The name also can be recalled by using the letters on the phone keypad. By contrast, standard memory dial recalls numbers from number-only locations.

AM (Amplitude Modulation) CW modulation using amplitude variation in proportion to the amplitude of the modulating signal; usually taken as DSB-LC for commercial broadcast transmissions and DSB-SC for multiplexed systems.

AMPS (Advanced Mobile Phone Service)   Analog cellular standard operating in the frequency range of 800 MHz with a bandwidth of 30kHz.

Analog   The traditional method of modulating radio signals so that they can carry information. AM (amplitude modulation) and FM (frequency modulation) are the two most common methods of analog modulation. Though most U.S. cellular systems today carry phone conversations using digital, many still offer analog transmission, especially in roaming situations.

ANSI (The American National Standards Institute)   A non-profit, privately funded membership organization that coordinates the development of U.S. voluntary national standards and is the U.S. representative to non-treaty international standards-setting entities, including the International Organization for Standardization (ISO).

Antenna   A device for transmitting and/or receiving signals. The size and shape of antennas are determined, in large part, by the frequency of the signal they are receiving. Antennas are needed on both the wireless handset and the base station.

API (Application Programming Interface)   Software used by an application program to request and carry out lower-level services performed by a computer's or telephone system's operating system.

ARDIS (Advanced Radio Data Information System)   A nationwide, public two-way wireless packet data network, originally developed by IBM, now owned and operated by American Mobile.

ASCII (American Standard Code for Information Interchange)   A standard code used by computer and data communication systems for translating characters, numbers, and punctuation into digital form. ASCII characters can be recognized by computer and communications devices using a variety of applications.

ATM (Asynchronous Transfer Mode)   A very high-speed transmission technology. ATM is a high-bandwidth, low-delay, connection-oriented switching and multiplexing technique. There are efforts underway to develop wireless ATM networks.

Authentication   A process used by the wireless carriers to verify the identity of a mobile station.

AVL (Automatic Vehicle Location)   The ability to pinpoint the location of a vehicle within a given range.

Bandwidth   Measure of the carrying capacity or size of a communication channel. For an analog circuit, the bandwidth is the difference between the highest and lowest frequencies that a medium can transmit and is expressed in hertz (Hz). Hz is equal to one cycle per second.

Baseband Frequency band occupied by information-bearing signals before combining with a carrier in the modulation process.

Base Station The central radio transmitter/receiver that maintains communications with mobile radiotelephones within a given range (typically a cell site).

Bluetooth A technology specification for small form factor, low-cost, short-range radio links between mobile PCs, mobile phones, and other portable devices. It is expected to enable users to connect a wide range of computing and telecommunications devices without the need to connect cables.

BPS (Bits Per Second) The unit of measurement for the rate at which data is transmitted.

Broadband Also called wideband. Transmission facility whose bandwidth is greater than that available on voice-grade facilities.

Browser Software that moves documents on the World Wide Web to your computer, PDA, or phone.

BTA (Basic Trading Area) A service area designed by Rand McNally and adopted by the FCC to promote the rapid deployment and ubiquitous coverage of Personal Communications Services (PCS). Built from county boundaries, BTAs generally cover a city and its surrounding environs. BTAs are component parts of Major Trading Areas (MTAs). There are 493 BTAs in the U.S.

CAI (Common Air Interface) A standard for the interface between a radio network and equipment. A CAI enables multiple vendors to develop equipment that will interoperate.

CDMA (Code Division Multiple Access) Spread Spectrum method of enabling multiple users to share the radio frequency spectrum by assigning each active user an individual code.

CDPD (Cellular Digital Packet Data) Technology that enables data files to be broken into a number of packets and be sent along idle channels of existing cellular voice networks. It is a wide area data network that takes advantage of exciting AMPS (US) cellular network by transmitting data packets on unused voice channels. Data is transmitted at an effective data rate of 14.4–19.2 Kbps.

Cell The basic geographic unit of a wireless system, also the basis for the generic industry term cellular. A city or county is divided into smaller cells, each of which is equipped with a low-powered radio transmitter/receiver. The cells can vary in size depending upon terrain, capacity demands, and so on. By controlling the transmission power, the radio frequencies assigned to one cell can be limited to the boundaries of that cell. When a wireless phone moves from one cell toward another, a computer at the Mobile Telephone Switching Office (MTSO) monitors the movement and, at the proper time, transfers (or hands off) the phone call to the new cell and another radio frequency. The hand-off is performed so quickly that it's unnoticeable to the callers.

Cell Site   The location at which communication equipment is located for each cell. A cell site includes antennas, a support structure for those antennas, and communications equipment to connect the site to the rest of the wireless system. This equipment is normally housed in a small shelter at the base of the site. Although many antennas are placed on towers, where existing structures provide for a site that is higher than its surroundings, antennas will be placed on them. For example, antennas have been placed on water towers, grain silos, and building rooftops.

Cellular Handoff   In cellular communication, a telephone call is switched by computers from one transmitter to the next, without disconnecting the signal, as a vehicle moves from cell to cell. The mobile remains on a specific channel until the signal strength diminishes and then is automatically told to go to another channel and pick up the transferred transmissions there.

Channel   A path along which a communications signal is transmitted.

Circuit Switched   A switching technique that establishes a dedicated and uninterrupted connection between the sender and the receiver.

Client-Server   A computer network system in which programs and information reside on the server and clients connect to the server for network access.

CMOS (Complementary metal oxide semiconductor)   A semiconductor technology chosen for its low-power consumption and good noise immunity.

CMRS (Commercial Mobile Radio Service)   The regulatory classification that the FCC uses to govern all commercial wireless service providers, including personal communications services, cellular, and Enhanced Specialized Mobile Radio.

Codec   Short for "coder and decoder," also COFIDEC (COder-FIlter-DECoder). Translates audio to digital signals and digital back to audio signals. This is usually accomplished with an A/D (analog to digital) and D/A (digital to analog) converter.

Co-Location   The site of two or more separate companies' wireless antennas on the same support structure.

Compression   Reducing the size of data to be stored or transmitted in order to save transmission time, capacity, or storage space.

COP (Computer operating properly)   This circuit is used to detect device runaway and provide a means for restoring correct operation.

CSMA/CA (Carrier Sense Multiple Access with Collision Avoidance)   A medium access control technique for multiple-access transmission media. A station wanting to transmit first senses the medium and transmits only if the medium is idle.

CT1 Cordless Telephone (first generation)   Analog cordless telephone standard operating at a frequency range of 46 to 49 MHz, with a bandwidth of 25 kHz.

CT2 Cordless Telephone (second generation)   System based on ETSI standard MPT1375 that describes a residential cordless phone operating at 864 to 886 MHz that can be used commercially.

CUG (Closed User Group)   Selected user groups that communicate freely within the group but have restricted incoming, and often outgoing, communications.

D-AMPS (Digital Advanced Mobile Phone Service)   A North American term for digital cellular radio.

DBS (Direct Broadcast Satellite)   A high-powered satellite, or satellite service, which sends signals to relatively small dishes installed at homes and office buildings.

DECT (Digital European Cordless Telephone)   Standard based on a micro-cellular radio based on a Time Division Multiple Access system that provides low-power cordless access between the subscriber and the base station up to a few hundred meters.

Demodulation   Process of recovering a low-frequency signal from a modulated carrier. Examples of low-frequency signals are voice or low-speed data.

DES (Data Encryption Standard)   A 56-bit, private key, symmetric cryptographic algorithm for the protection of unclassified computer data developed by IBM in 1977.

Dial-Up   The use of a standard telephone to create a telephone or data call.

Digital Modulation   A method of encoding information for transmission. Information (in most cases, a voice conversation) is turned into a series of digital bits, the 0s and 1s of computer binary language. At the receiving end, the information is reconverted to its original form. Digital transmission offers a cleaner signal and is less immune to the problems of analog modulation, such as fading and static. *See also* CDMA, TDMA, and GSM.

Digital Telephone   Telephone terminal that digitizes voice signals and DTMF tones for transmission over a regular twisted pair of copper to the CO/PBX. The process is reversed on the way back from the Central Office/PBX.

DSSS (Direct Sequence Spread Spectrum)   This generates a redundant bit pattern for each bit to be transmitted. This bit pattern is called a chip (or chipping code). The longer the chip, the greater the probability that the original data can be recovered (and, of course, the more bandwidth required). Even if one or more bits in the chip is damaged during transmission, statistical techniques embedded in the radio can recover the original data without the need for retransmission. To an unintended receiver, DSSS appears as low-power wideband noise and is rejected (ignored) by most narrowband receivers.

Down converter   Usually an integrated device that provides gain and frequency translation to a lower frequency.

DSP (Digital Signal Processor)   A specialized computer chip that performs calculations on digitized signals that were originally analog and then sends the results.

DTMF (Dual Tone Multi-Frequency)   Tone-dialing system based on outputting two non-harmonic-related frequencies simultaneously to identify the number dialed. Eight frequencies have been assigned to the four rows and four columns of a typical keypad.

Duplex   Mode of operation permitting the simultaneous transmission and reception of signals.

E-mail (Electronic Mail)   Messages sent across communication networks, both wireless and landline.

EMC (Electromagnetic Compatibility)   The capability of equipment or systems to be used in their intended environment within the designed efficiency levels without causing or receiving degradation due to unintentional electromagnetic interference. The proper shielding of devices reduces interference.

Encryption   The transformation of data, for the purpose of privacy, into an unreadable format until reformatted with a decryption key. Public key encryption utilizes the RSA encryption key (which stands for its developers, Rivest, Shamir, and Adleman). PGP, or Pretty Good Privacy, is a cryptography program for computer data, e-mail, and voice conversation.

EPROM (Erasable programmable read only memory)   This type of memory requires exposure to ultra-violet wavelengths in order to erase previous data.

Equalizer   Electrical network in which phase delay or gain varies with frequency to compensate for an undesired amplitude or phase characteristic in a frequency-dependent transmission line.

ESMR (Enhanced Specialized Mobile Radio)   Digital mobile telephone services offered to the public over channels previously used for two-way analog dispatch services.

ESN (Electronic Serial Number)   The unique number assigned to a wireless phone by the manufacturer. According to the Federal Communications Commission, the ESN is to be fixed and unchangeable, a sort of unique fingerprint for each phone. *See also* MIN.

ETACS (Extended Total Access Communications Systems)   The conventional wireless technology used in the United Kingdom and other countries. It was developed from the U.S. AMPS technology.

Extranet   An Intranet-like network that a company extends to conduct business with its customers and/or its suppliers. Extranets generally have secure areas to provide information to customers and external partners.

FCC (Federal Communications Commission)   The government agency responsible for regulating telecommunications in the U.S.

FDMA (Frequency Division Multiple Access)   Method of radio transmission that enables multiple users to access a group of radio frequency bands without interference by assigning each active user an individual frequency channel.

FHSS (Frequency Hopping Spread Spectrum)   A technique used in spread spectrum radio transmission systems, such as wireless LANs and some PCS cellular systems, that involves the conversion of a datastream into a stream of packets.

FM (Frequency Modulation)   CW modulation using frequency variation in proportion to the amplitude of the modulating signal.

Frequency   A measure of the energy, as one or more waves per second, in an electrical or light-wave information signal. A signal's frequency is stated in either cycles per second or Hertz (Hz).

Frequency Hopping   In this type of spread spectrum approach, both units (base and subscriber or handset and base) hop from frequency to frequency in a simultaneous fashion.

Frequency Re-Use   The ability to use the same frequencies repeatedly within a single system, made possible by the basic design approach for wireless. Since each cell is designed to use radio frequencies only within its boundaries, the same frequencies can be reused in other cells not far away with little potential for interference.

GHz (GigaHertz [Billions of Hertz])   Personal Communications Services operate in the 1.9-GHz band of the electromagnetic spectrum. *See also* Hertz, KHz, MHz.

GPRS (General Packet Radio Service)   An extension to the GSM standard to include packet data services. It is expected to be launched in 2001.

GPS (Global Positioning System)   A satellite system using 24 satellites orbiting the earth at 10,900 miles that enables users to pinpoint precise locations using the satellites as reference points.

GSM (Global System for Mobile Communications)   A world standard for digital wireless transmissions. GSM is the most widely used standard in the world today with more than 150 million users worldwide. *See also* TDMA.

GSO (Geosynchronous Satellite Orbit)   A satellite in orbit 23,000 miles over the equator with an orbit time of 24 hours. Also known as geostationary.

GUI (Graphical User Interface)   A name for any computer interface that substitutes graphics for characters.

Guyed   A type of wireless transmission tower that is supported by thin guy wires. *See also* Monopole.

Hand-Off   The process by which the Mobile Telephone Switching Office (MTSO) passes a wireless phone conversation from one radio frequency in one cell to another radio frequency in another cell. It is performed quickly enough that callers don't notice.

Hands-Free   A feature that permits a driver to use a wireless car phone without lifting or holding the handset. An important safety feature.

HDML (Handheld Device Markup Language)   A modification of standard HTML, developed by Unwired Planet, for use on small screens of mobile phones, PDAs, and pagers. HDML is a text-based markup language that uses the HyperText Transfer Protocol (HTTP) and is compatible with Web servers.

Hertz   A measurement of electromagnetic energy, equivalent to one wave or cycle per second.

HTML (HyperText Markup Language)   An authoring software language used on the Web. HTML is used to create Web pages and hyperlinks.

HTTP (HyperText Transfer Protocol)   The protocol used by the Web server and the client browser to communicate and move documents around the Internet.

iDEN (Integrated Dispatch Enhanced Network)   A wireless technology developed by Motorola that works in the 800-MHz, 900-MHz, and 1.5-GHz radio bands. The technology supports on one handset voice, both dispatch radio and it uses PSTN connection-numeric paging, Short Message Service (SMS), and data and fax transmission.

IEEE 802.11   Specification for Wireless Local Area Networks from the 802.11 committee of the Institute of Electrical and Electronic Engineers (IEEE). It is a standard for 1- and 2-Mbps wireless LANs and has a single MAC layer for the following physical-layer technologies: Frequency Hopping Spread Spectrum, Direct Sequence Spread Spectrum, and Infrared.

IMSI (International Mobile Station Identifier)   A number assigned to a mobile station by the wireless carrier uniquely identifying the mobile station nationally and internationally.

IMT-2000 (International Mobile Telecommunications-2000)   The standard for third-generation (3G) mobile communications systems. In Europe, it is called UMTS and in Japan it is called J-FPLMTS.

Independent network (stand-alone network)   A network that provides (usually temporarily) peer-to-peer connectivity without relying on a complete network infrastructure.

Infrared   A band of the electromagnetic spectrum used for airwave communications and some fiber-optic transmission systems. Infrared is commonly used for short-range (up to 20 feet), through-the-air data transmission. Many PC devices have infrared ports, called Infrared Serial Data Links (IRDA), to synchronize with other devices. IRDA supports speeds up to 1.5 Mbps.

Interconnection   The routing of telecommunications traffic between the networks of different communications companies.

Internet A global network of linked computer networks made user-friendly, thus popularized, by a graphical interface called the World Wide Web.

Intranet An internal network, which is private or employs a firewall to secure it from outside access that supports Internet technology. The Intranet is used for inter-company communications and can be accessed only by authorized users.

IP (Internet Protocol) See TCP/IP.

ISM Industrial, Science, and Medical Bands of frequencies that are allocated by the FCC to spur the rapid development of RF applications in a virtual open-market fashion. Licensing is automatic.

IXC (Interexchange Carrier) A long-distance phone company.

Java A programming language from Sun Microsystems that abstracts data on bytecodes so that the same code runs on any operating system. Java software is generally posted on the Web and is downloadable over the Internet to a PC. HotJava is installed on a Web browser and enables Java programs to be delivered over the Web and run on a PC.

Jini A technology from Sun Microsystems that is expected to enable devices to link together to form an ad hoc community, without installation or human intervention.

J-FPLMTS (Japanese Future Public Land Mobile Telecommunications Services) The Japanese equivalent of the IMT-2000 third-generation (3G) technology standard.

KHz (KiloHertz [Thousands of Hertz]) Each wireless phone call occupies only a few KHz.

LAN (Local Area Network) A data communications network, typically within a building or campus, that links computers and peripheral devices under some form of standard control.

LEC (Local Exchange Company) The traditional, local, wired monopoly phone company.

LEO (Low Earth Orbit) An orbital plane a few hundred miles above the earth. A new generation of communication satellites is being launched in this orbit. LEO satellites are generally divided into two groups: big and little LEOs, with each group assigned specific radio frequencies. Big LEOs support both voice and data communications, whereas little LEOs support only data communications.

Limiter Circuit whose output signal amplitude remains at some predetermined level in spite of wide variations in input signal amplitude.

LMDS (Local Multipoint Distribution System) A system developed by Bellcore for Wireless Local Loop (WLL) applications. In the U.S., the FCC set aside a total LMDS bandwidth of 1.15 GHz in the 28-GHz, 30-GHz, and 31-GHz frequency bands. LMDS supports voice and high-speed interactive data, with the potential to provide bandwidth of as much as 500 Mbps.

LNA Low Noise Amplifier Usually the first active gain device in a receiver, its purpose is to provide amplification to a low-level signal from a large number of available signals while minimizing noise.

MAN (Metropolitan Area Network) A network covering a larger area than a Local Area Network (LAN) and less than a Wide Area Network (WAN). Typically, a MAN connects two or more LANs. In addition to data, a MAN may also carry voice, video, image, and multimedia.

Message Alert A light or other indicator on a wireless phone that notifies a user that a call has come in. This is a useful feature, especially if the wireless subscriber has voice mail. This is also called a call-in-absence indicator.

MHz (MegaHertz) Millions of Hertz Cellular and ESMR systems operate in the 800- and 900-MHz bands of the electromagnetic spectrum. *See also* Hertz, KHz, GHz.

MIN (Mobile Identification Number) A number assigned by the wireless carrier to a customer's phone. The MIN is meant to be changeable, since the phone could change hands or a customer could move to another city. *See also* ESN, IMSI, TMSI.

MIME (Multipurpose Internet Mail Extensions) The standard format, developed and adopted by the Internet Engineering Task Force (IETF), for including non-text information in Internet mail, thus supporting the transmission of mixed-media messages across TCP/IP networks. In addition to covering binary, audio, and video data, MIME is the standard for transmitting foreign language text that cannot be represented in ASCII code.

Mixer Device that utilizes its non-linear characteristics to provide frequency conversions from one frequency to another. This may be from a relatively high frequency to an intermediate frequency (IF). In this case, it is known as a down-mixer. Or it may be from a lower frequency to a higher frequency, such as the carrier frequency, for example. In this case, it is known as an up mixer.

MMDS (Microwave Multi-point Distribution System or Multipoint Multichannel Distribution Service) A method of distributing cable television signals through microwave from a single transmission point to multiple receiving points.

Mobitex   A cellular land radio-based packet switched data communications system used by BellSouth's two-way packet data network and developed by Ericsson.

Modem (MOdulator-DEModulator)   Unit that modulates and demodulates digital information from a terminal or computer port to an analog carrier signal for passage over an analog line.

Monopole   A slender self-supporting tower on which wireless antennas can be placed. *See also* Guyed.

MSA (Metropolitan Statistical Area)   An MSA denotes one of the 306 largest urban population markets as designated by the U.S. government. Two cellular operators are licensed in each MSA.

MSS (Mobile Satellite Service)   Communications transmission service provided by satellites. A single satellite can provide coverage to the entire U.S.

MTA (Major Trading Area)   A service area designed by Rand McNally and adopted by the FCC to promote the rapid deployment and ubiquitous coverage of Personal Communications Services (PCS). Built from Basic Trading Areas (BTAs), MTAs are centered on a major city and generally cover an area the size of a state. There are 51 MTAs in the United States. *See also* BTA.

MTSO (Mobile Telephone Switching Office)   Name given to the central coordinating element. Designated as common control, the MTSO is made up of signal processors, memories, switching networks, trunk circuits, and ancillary services.

NAM (Number Assignment Module)   The NAM is the electronic memory in the wireless phone that stores the telephone number and electronic serial number.

Network Computer   An inexpensive (approximately $500) personal computer that does not have a hard disk, but it can be used to browse the Internet and run applications on a server on the Internet or corporate Intranet. The concept of the network computer, an inexpensive easy-to-use tool for the masses, is promoted by Oracle.

Nondirectional Antenna   An antenna that transmits and receives equally well in all directions, usually on one plane. Also called an omnidirectional antenna.

Non-ionizing Emissions   Radio waves, infrared rays, and visible light rays, none of which can affect an atom's electrical balance.

Non-Wireline Cellular Company   The Block A carrier. The A originally stood for alternate, that is, the non-Bell or B carrier in a market. The FCC, in setting up the licensing and regulatory rules for cellular, decided to license two cellular systems in each market. It reserved one for the local telephone company and opened a second system, the Block A system, to other interested applicants. The distinction between Block A and Block B was meaningful only during the licensing phase at the FCC. Once a system is constructed, it can be sold to anyone. Thus, in some markets today, both the A and B systems are owned by telephone companies; one happens to be the local phone company for the area and the other is a phone company that decided to buy a cellular system outside its home territory. *See also* Wireline Cellular Company.

Operating System   A software program that manages the basic operations of a computer system. These operations include memory apportionment, the order and method of handling tasks, the flow of information into and out of the main processor and to peripherals, and so on. Companies involved in wireless data operating systems include Microsoft and Symbian.

PA Power Amplifier   Provides the high power gain to the transmitter. Typical figures of merit include gain, efficiency, linearity (in amplitude and phase-modulated systems), and stability.

Packet   A bundle of data organized in a specific way for transmission. The three principal elements of a packet include the header, the text, and the trailer (error detection and correction bits).

Packet Radio   The transmission of data over radio using a version of the X.25 data communications protocol. The data is broken into packets and is transmitted wirelessly.

Packet Switching   Sending data in packets through a network to a remote location. The data sent is assembled by the PAD (see definition), often called a modem, into individual packets of data.

PAD (Packet Assembler/Disassembler)   A device that assembles characters into packets that are transmitted by a packet switching network. A PAD also receives packets and disassembles them into a format that can be handled by the terminal or host.

Pager   Small portable receivers that are generally inexpensive, reliable, and have nationwide coverage. Pagers began as one-way devices, but two-way paging capabilities are available over some networks, notably packet data and narrowband PCS networks.

PC Card   The new name for PCMCIA cards (see definition). A small, credit-card sized device, compatible with the PCMCIA PC Card standard, that packages for memory and input/output.

PCMCIA (Personal Computer Memory Card International Association)   A standards body that sets the standards for PC cards.

PCS (Personal Communications Services)   A service that bundles voice communications, numeric and text messaging, voice-mail, and other features into one device, service, or bill.

PDA (Personal Digital Assistant)   Portable computing devices capable of transmitting data. These devices make possible services such as paging, data messaging, e-mail, stock quotations, handwriting recognition, personal computing, facsimile, date book, and other information-handling capabilities.

PHS (Personal Handyphone System)   System developed in Japan to be the 1.9-GHz residential cordless/wireless PBX system.

Picocell   A wireless base station with extremely low output power designed to cover an extremely small area, such as one floor of an office building.

PIM (Personal Information Manager)   Also known as a contact manager, this is a form of software that logs personal and business information, such as contacts, appointments, lists, notes, occasions, and so on.

PLL (Phase Locked Loop)   PLL is a major component in the frequency synthesizer scheme. This device provides a wide, flexible range of internal frequency dividers that enable the designer the ability to create a synthesizer to match design requirements.

POP   Short for population. One POP equals one person. For example, a carrier whose market serves one million people is said to offer service to one million POPs. In the wireless industry, systems are valued financially based on the population of the market served.

Portable   Portable phones are small, handheld units that can fit in a pocket, briefcase, or purse. Using an attachment, many can be plugged into an automobile cigarette lighter to save battery power. As a smaller, lighter phone, a portable operates at power levels of up to 6/10ths of a watt. Furthermore, digital phones are almost always portable phones. *See also* Mobile, Transportable.

POS (Point-of-Sale Terminal)   A type of computer terminal used to collect and store retail sales data. Wireless POS terminals are often used for remote and temporary locations.

Prescaler Device that divides down the high frequency sample of the VCO to a frequency within the range of operation of the PLL. Most prescalers are capable of operating at different divide ratios. The most popular is known as dual modulus.

Protocol A specific set of rules for organizing the transmission of data in a network.

QPSK (Quadrature Phase Shift Keying) Spectrally efficient modulation technique that breaks the information path into two parts, the "in phase" and the "quadrature phase" components. The combination of these two signals creates one of four unique symbols that are then used to modulate the phase of the carrier.

Receiver Arrangement of active components such as the LNA and IF amplifier together with passive components such as the image filter and IF filter. Taken together they perform the task of recovering the modulation from a known RF signal while rejecting unwanted signals. This is the portion of the communication system that includes a detector and signal-processing electronics to convert electrical signals (electric waves) to audio or data signals. It provides reception and, if necessary, demodulation of electronic signals.

Repeater Amplifier and the associated equipment used in a telephone circuit to process a signal and retransmit it.

RF (Radio Frequency) Frequencies of the electromagnetic spectrum normally associated with radio wave propagation. Sometimes defined as a transmission at any frequency at which coherent electromagnetic energy radiation is possible.

Roaming The ability to use a wireless phone to make and receive calls in places outside one's home calling area. Wireless communication is limited by how far signals carry for given power output. Wireless LANs use cells, called microcells, similar to the cellular telephone system to extend the range of wireless connectivity. At any point in time, a mobile PC equipped with a Wireless LAN adapter is associated with a single access point and its microcell, or area of coverage. Individual microcells overlap to enable continuous communication within a wired network. They handle low-power signals and hand off users as they roam through a given geographic area.

RSA (Rural Service Area) One of the 428 FCC-designated rural markets across the U.S. Two cellular carriers are licensed in each RSA. *See also* MSA, CGSA.

Sensitivity For a receiver, the input signal (in uV or mV) required for a specific output level.

Service Charge   The amount paid each month to receive wireless service. This amount is fixed and is to be paid regardless of how much or how little the wireless phone is used.

SIM (Subscriber Identity Module)   A computer chip set in a handset that contains information needed to identify the subscriber when connecting to the network, especially for billing purposes.

Smart Card   A credit card-sized card with a microprocessor and memory.

Smart Phone   A phone with a microprocessor, memory, screen, and a built-in modem. The smart phone combines the some of the capabilities of a PC on a handset.

SMR (Specialized Mobile Radio)   Private business service using mobile radiotelephones and base stations similar to other wireless services. It is usually used in dispatch applications, such as delivery companies or taxicab organizations. SMR is the forerunner of ESMR service. *See also* ESMR.

SMS (Short Message Service)   A service to send short alphanumeric messages between devices.

Spectrum   Most often used in the context of frequency allocations, this refers to the frequencies allowed for a type of service out of the total available.

Spoofing   An access method that supports a very fast dial-up routine in a switched network that mimics the functionality of a packet switched data network.

Spread Spectrum   A technique to reduce and avoid interference by taking advantage of the statistical means to send a signal between two points. Spread Spectrum is a modulation technique, also known as frequency hopping, used in wireless systems. The data is packetized and spread over a range of bandwidth.

Standby Time   The amount of time a fully charged wireless portable or transportable phone can be on before the phone's battery will lose power. *See also* Talk Time.

Store-and-Forward   The ability to transmit a message to an intermediate relay point and to store it temporarily when the receiving device is unavailable.

Synchronization   Also known as replication, it is the process of uploading and downloading information from two or more databases, so that each is identical.

Talk Time   The length of time one can talk on a portable or transportable wireless phone without recharging the battery. The battery capacity of a phone is usually expressed in terms of minutes of talk time or hours of standby time. When one is talking, the phone draws more power from the battery. *See also* Standy Time.

TCP/IP (Transmission Control Protocol/Internet Protocol)   The standard set of protocols used by the Internet for transferring information between computers, handsets, and other devices.

TDMA (Time Division Multiple Access)   Technique that assigns each subscriber desiring service a different time slot on a given frequency. Signal compression is achieved by running at very high frequencies. Each user can then deliver the fixed packet message in a brief burst of time, thereby increasing the capacity of the system.

Telecommunications Act of 1996   Signed into law by President Clinton on February 8, 1996, it establishes a pro-competitive, deregulatory framework for telecommunications in the U.S.

Telemetry   A wireless or landline system for the transmission of data (either digital or analog) for remote monitoring.

Third Generation   See 3G.

TMSI (Temporary Mobile Station Identifier)   A mobile station identifier (MSID) sent over the air interface and is assigned dynamically by the network to the mobile station.

Transmitter   Equipment that feeds the radio signal to an antenna for transmission. It consists of active components such as the up mixer, driver, PA, and passive components such as the TX filter. Taken together, these components impress a signal onto an RF carrier of the correct frequency by instantaneously adjusting its phase, frequency, or amplitude and provide enough gain to the signal to project it through the ether to its intended target.

Transportable Phone   Transportable phones, or bag phones are essentially car phones with the handset, antenna, and battery packaged together in a carrying case. They can be plugged into a car's cigarette lighter or can operate off of a portable battery pack for use anywhere. Like a mobile phone, transportable phones can operate at up to three watts of power. Although technically portable, the transportable should not be confused with the handheld, one-piece wireless phone. *See also* Portable, Mobile.

Trunk   Telephone circuit or channel between two central offices or switching entities.

UMTS (Universal Mobile Telecommunications Services) The European term for wireless systems based on the IMT-2000 standard.

Upconverter Usually an integrated device that includes an up mixer and some amplification. Other functions it might perform are power control and/or transmit envelope shaping.

VCO (Voltage Controlled Oscillator) Oscillator whose output frequency varies with an applied dc control voltage.

Vocoder (Voice Coder) A device used to convert speech into digital signals. Compresses the voice to be transmitted and then expands the voice received in digital systems. A variety of algorithms can accomplish this.

Voice-Activated Dialing A feature that permits one to dial a phone number by speaking to a wireless phone instead of using a keypad. The feature contributes to convenience as well as driving safety.

Voice Mail A computerized answering service that answers a call, plays a greeting, and records a message. Depending on the sophistication of the service, it also can notify the subscriber, via pager, that he or she has received a call. This is also called voice messaging. *See also* LAN, MAN.

WAN (Wide Area Network) A network that uses local telephone company lines to connect geographically dispersed sites.

WAP (Wireless Applications Protocol) A proposed protocol for wireless applications. The protocol is designed to simplify how wireless users access electronic and voice mail, send and receive faxes, make stock trades, conduct banking transactions, and view miniature Web pages on a small screen.

Wireless A radio-based system that enables the transmission of telephone and/or data signals through the air without a physical connection, such as a metal wire or fiber optic cable.

Wireless Node A user computer with a wireless network interface card (adapter).

Wireline Cellular Carrier The Block B carrier. Under the FCC's initial cellular licensing procedures, the Block B carrier is the local telephone company's licensee. The FCC reserved one of the two systems in every cellular market for the local telephone (or wireline) company. With initial licensing complete, the distinction has slowly disappeared. The local phone company can sell its cellular system to anyone. *See also* Non-Wireline Cellular Carrier.

WLL (Wireless Local Loop)   A local wireless communications network that bypasses the local exchange carrier and provides high-speed, fixed data transmission.

WML (Wireless Markup Language)   A compact version of the Handheld Device Markup Language. *See* HDML.

WRC (World Radio Conference)   Formerly known as WARC, or World Administrative Radio Conference, it is an international conference that sets international frequencies.

X.25   A communication protocol for packet switched public data networks.

INDEX

Symbols

3G (third generation) GSM, terminals, 158
3G wireless, 135
3G/UMTS, 305
5-UP (Unified Protocol), 247
8-PSK (Phase Shift Keying), EDGE, 304
802.11 specifications, WLANs, 232–235, 240
802.11a standard, 244–245

A

Abis interface, GSM, 166
AC (Alternating Current), 42
ACA (Automatic Channel Assignment), TDMA
    cellular, 91
academic wireless applications, 343–344
access grant channels, GSM, 164
access points
    wireless systems, 241
    WLANs, 227
Accompli (Motorola), 258, 350
ACK (acknowledgement character), WLANs, 236
action cams, 52
ACTS satellite systems, 62
adapters, wireless systems, 242
air interfaces
    CDMA, 101
    GSM, 161–162
Alcatel wireless applications, 325
allocating frequency, cellular, 125
AM (Amplitude Modulation), 34
amplitude, sound waveforms, 8
AMPS (Advanced Mobile Phone Service), 13,
    24, 190
analog cellular systems, 108–111
    churn, 110
    FDMA, 109

analog communications, 4
ANSI (American National Standards Institute),
    frequency allocation, 19
AOD (Audio on Demand), 331
apparatus types, wirelsss data, 216
applications, wireless. See wireless applications.
AT&T PocketNet, 258
ATMover Satellite, 270
attenuation, radio frequencies, 47
AuC (Authentication Center), GSM, 161
authentication, Over-the-Air Activation, 150
automatic roaming, digital cellular systems, 131

B

backbones, GPRS, 207
bandpass filters, 10
bandwidth, 53
    802.11a standard, 245
    EDGE efficiency, 305
    microwave communications, 53–55
    radio frequencies, 48
    satellite systems, 69
    spread spectrum, 97
    WLANs, 221
    WLL, 289
base stations
    digital cellular systems, 119
    GSM, 160
    LMDS, 294
bearer service, GPRS, 209
Bluetooth, 248–249
    data channels, 251
    GPRS, 303
    PANs, 250
    PDAs, 340
    piconets, 252
    voice channels, 251

BRAN (Broadband Radio Access Networks), 247
British Telecommunications wireless
      applications, 325
broadband, 256
  ATM over Satellite, 270
  competitive requirements, 294
  IPover Satellite, 269
  LMDS, 290
  Mobile IP, 267
  TCP over wireless, 266–267
  TCP/IP over Satellite, 268–269
  Teledesic, 273–278
broascast control channels, GSM, 164
BSCs (Base Station Controllers), GSM, 160
BSSMAP (Base Station System Mobile
      Application Part), GSM, 166
BTSs (Base Transceiver Stations), GSM, 160

C

C-band, satellite systems, 58
CAGR (Compound Annual Growth Rate), wireless
      data, 185
call setup, digital cellular systems, 116
CAPs (Competitive Access Providers), 51
CC (Control Channel), iDEN™, 177
CCK (Complementary Code Keying), 232
CDG (CDMA Development Group), 142
CDMA (Code Division Multiple Access), 30–31,
      95, 99
  air interfaces, 101
  channels, 101–105
  digital PCS, 142–149
  GPS, 100
  IWUs (Interworking Units), 150
  L interface, 151
  packet data services, 152
  pseudo-random code sequences, 97
  RLP (Radio Link Protocol), 150
  security, 100
  simultaneous voice/data transmission, 151
  standards, 319–320
  synchronization, 100
  transmission speeds, 101
  UMTS, 319
  vocoders, 100

CDPD (Cellular Digital Packet Data), 185,
      195–197
Celestri, 64–65
cell patterns, digital cellular systems, 121
cell phones, 134
cell site configuration, digital cellular
      systems, 122
cell splitting, digital cellular systems, 113
cellular modems, wireless data, 191–192
cellular systems
  AMPS, 24
  analog, 108, 111
  automatic roaming, 131
  CDMA, 95, 100, 142–148
    air interfaces, 101
    channels, 101–105
    GPS, 100
    pseudo-random code sequences, 97
    synchronization, 100
    transmission speeds, 101
    vocoders, 100
  churn, 110
  DCCH, 88
  digital systems. *See* digital systems.
  FDMA, 109
  handback, 129
  HLR, 127
  intersystem handoff, 128
  IS-136 TDMA standard, 87–92
  IS-41, 130
  landline to mobile calls, 126
  MSC, 126
  PCS user densities, 87
  PCS, 24–25
  SS7, 130
  standards, 136–137
  TACS, 24
  TDMA, 87
  third-party handoff, 129
  VLR, 127
cellular telephone calls, 13
CEPT (Conference of European Post and
      Telegraph), 156
channels, radio, 25
  AM, 34
  CDMA, 30–31
  DM, 36
  duplex separation, 26

FDMA, 28
FM, 35
modulation, 32–33
multiplexing, 28
TDMA, 29, 91
churn, 109–110
COFDM (Coded OFDM), 245
college wireless applications, 343–344
commercial wireless applications, 78
communications
microwave, 49–55
multiplexing, 28
radio channels, 25–36, 93
satellite, 56–65, 68–72, 75–77
configurations of WLANs, 228–229
connections
CDMA, 149
wireless data, 190, 193
conservation of energy, radio frequencies, 41
contention window, WLANs, 235
control channels
digital cellular systems, 115
GSM, 164
CRC (Cyclic Redundancy Check), WLANs, 238
CS-CDPD (Circuit-Switched Cellular), 195
CSMA/CA (Carrier Sense Multiple Access with Collision Avoidance), 234
CTIA (Cellular Telecommunications Industry Association), 19
custom data transport protocols, 266
custom TCP, 266

D

D-AMPS (Digital Advanced Mobile phone system), 305
data burst, GSM, 164
data channels, Bluetooth, 251
data rates, CDMA, 151
data services, TDMA cellular, 90
DBS (Direct Broadcast System), 77
DCA (Dynamic Channel Assignment), TDMA cellular, 91
DCCH (Digital Control Channel), 88
DCF (Distributed Coordination Function), 235
delay. *See* latency.

dial-up connections, 193
DIFS (Distributed Coordination Function Inter-Frame Spacing), 235
digital cellular systems, 24, 84, 112
automatic roaming, 131
base stations, 119
call setup, 116
cells, 113, 121–122
control channels, 115
DCCH, 88
failing signals, 115
frequency allocation, 125
frequency reuse, 120–121, 124
handback, 129
handoff, 117–118
HLR, 127
incoming call setup, 117
intersystem handoff, 128
IS-136 TDMA standard, 87–92
IS-41, 130
landline to mobile calls, 126
MSA, 113
MSC, 126
MTSO, 114, 119–120
overlapping coverage, 122
PBS user densities, 87
PCS, 25
sectorized cell coverage, 123
SS7, 130
TDMA, 87
third-party handoff, 129
tiered sites, 124
VLR, 127
Digital PCS, 137
CDMA, 142–149
Over-the-Air Activation, 149
TDMA, 139–140
transmission methods, 138
DM (Digital Modulation), 36
DAMPS (Digital AMPS), 25
down converters, 283
downlinks, satellite systems, 57
drumbeat signaling, 3
DS (Direct Sequence), spread spectrum, 96
DSI (Digital Speech Interpolation), ETDMA, 94
DSSS (Direct sequence spread spectrum), 142, 232
duplex separation, radio channels, 26

E

ECSA (Exchange Carriers Standards
    Association), frequency allocation, 21
EDACS (Enhanced Digital Access
    Communications System), 180
EDGE (Enhanced Data for Global Environment),
    93, 210, 298
  8-PSK, 304
  bandwidth efficiency, 305
Eggy, 327
EIA (Electronic Industries Association), 19
EIR (Equipment Inventory Register), GSM, 161
electrical signals, attenuation, 47
electromagnetic spectrum, 41–47
e-mail alerts, SMS, 203
energy, 40–41
enhanced voice services, GSM, 172–173
entropy, radio frequencies, 41
Ericsson EDACS (Enhanced Digital Access
    Communications System), 180
ETDMA (Extended TDMA), 94
ETSI (European Telecommunications Standards
    Institute), 156
explicit feedback, TCP over wireless, 267
extended WLANs, 229

F

failing signals, digital cellular systems, 115
Faraday Cage, 44
fast packet switching, Teledesic, 275
fax mail, SMS, 203
fax rates, CDMA, 151
FCA (Fixed Channel Assignment), TDMA
    cellular, 91
FCC (Federal Communications Commission),
    frequency registration, 7
FDMA (Frequency Division Multiple Access), 28,
    82–83, 109
FH (Frequency Hopping), spread spectrum, 96
fixed wireless, 337
FM (Frequency Modulation), 35
footprint, satellite systems, 57
forward channel, CDMA, 101

FRA. *See* WLL.
frames, TDMA, data burst, 164
free space communications, 5
frequencies
  802.11a standard, 245
  radio, 44–49
  wavelength relationship, 47
frequency allocation
  ANSI, 19
  digital cellular systems, 125
  ECSA, 21
  spectrum usage, 22
  standards groups, 19
  TIA, 19
frequency bands, 57, 281
frequency control channels, GSM, 164
frequency hopping, Geotek, 180
frequency reuse, digital cellular systems,
    120–121, 124
frequency spectrum, 7–9
FSK (Frequency Shift Keying), digital
    modulation, 36

G

GAIT (GSM/ANSI-136 Interoperability
    Team), 169
gamma rays, 47
gateways, Ricochet, 262
GEO satellites (geosynchronous), 17–18, 59,
    64–65, 269
Geotek, frequency hopping, 180
GGSNs (Gateway GPRS Support Nodes), 207
GMSK (Gaussian Minimum Shift Keying),
    197, 304
GPRS (General Packet Radio Services), 135,
    173, 298
  Bluetooth, 303
  GGSNs (Gateway GPRS Support Nodes), 207
  GSNs (GPRS Support Nodes), 206
  GTP (GPRS Tunneling Protocol), 301
  parallel GSM services, 210
  SGSNs (Serving GPRS Support Nodes), 206
  short and bursty transmissions, 300
  wireless Internet, 206–209
  X.25 support, 301

GPS (Global Positioning System), 77, 100
GSM (Global System for Mobile
    Communications), 137, 156, 159
    Abis interface, 166
    air interfaces, 161–162
    AuC, 161
    BSCs, 160
    BSSMAP, 166
    BTSs, 160
    channels, 163–164
    EDGE, 92
    EIR, 161
    enhanced voice services, 172–173
    HLR, 161
    HSCSD, 305
    IIF, 170
    link interfaces, 161
    LPC, 165
    MAP, 170
    mobile units, 159
    MSCs, 160
    MTP, 166
    network, 160, 165, 170
    parallel GPRS services, 210
    SCCP, 166
    SIMs, 159–160, 317
    speech coding formats, 165
    subscriber base, 157
    TCHs, 163
    TDMA frames, data burst, 164
    third generation, 158
    VLR, 161
    WLL, 173
GSNs (GPRS support nodes), 206
GTP (GPRS Tunneling Protocol), 301

H

half-rate channels, GSM, 163
handback, digital cellular systems, 129
handheld Internet devices, wireless Internet, 313
handoff, digital cellular systems, 117–118
handsets, wireless Internet, 326–329
hardware developers, wirelsss data, 215
HDML (Handheld Device Markup
    Language), 258

HF (High Frequency) band, 11
HiperLAN/2, 247
HLR (Home Location Register)
    digital cellular systems, 127
    GSM, 161
HomeRF, 238–240
HSCSD (High-Speed Circuit-Switched Data), 305

I

IDEN™, 173–175
    CC (Control Channel), 177
    licensing blocks, 178
    multiplied channel capacity, 175
    service areas, 178
    speed of calls, 179
    VSELP vocoder, 174–175
IEEE 802.11b standard, HomeRF, 238
IETF (Internet Engineering Task Force), 271
IIF (Interworking and Interoperability Function),
    GSM, 170
IMEIs (International Mobile Equipment
    Identities), GSM mobile units, 160
IMSIs (International Mobile Subscriber
    Identities), GSM SIMs, 160
IMT-2000 initiative, 305–306
IMTS (Improved Mobile Telephone Service), 13
incoming call setup, digital cellular systems, 117
independent WLANs, 228
industry occupations for wirelsss data, 214
infrared WLANs
    links, 223
    transmissions, 234
infrastructure WLANs, 229
integrated services, 175
integrity of WLANs, 231
interfaces, GSM, 161
Internet
    IP, 271
    mobility. See mobile Internet.
    standards groups, 270
    wireless data, 203–209
intersystem handoff, digital cellular systems, 128
interworking, SMS, 201
IP (Internet Protocol), 271
    GPRS support, 301
    UMTS compatibility, 318

IP over Satellite, 269
IS-136 TDMA standard, 87–92
IS-41 digital cellular systems, 130
ISPs (Internet Service Providers), satellite
   links, 333
IWF (Interworking Function), 267
IWUs (Interworking Units), CDMA, 150

J–K

jamming microwave transmissions, 13
Ka-band, 58, 276
kinetic energy, 41
Ku-band, satellite systems, 58
Kyocera PCS PDA phones, 349

L

L interface, CDMA, 151
landline-to-mobile calls, 126
LANs, wireless. *See* WLANs.
Lap Link™, 257
laptops, 257
latency
   satellite systems, 64, 68–69, 272, 335–336
   TCP, 272
   Teledesic, 277
leased lines, radio-based systems, 5
LEO (Low Earth Orbit) satellites, 60, 69–72
   advantages, 75–76
   GEO hybrids, 65
   latency, 272
licensing blocks, iDEN™, 178
light, 37, 46
line of sight, microwave communications, 4, 55
link interfaces, GSM, 161
LMDS (Local Multipoint Distribution
      Services), 290, 294–295
location services, 351
LPC (Linear Predictive Coding), GSM, 165

M

MAC layer, WLANs, 234
make before break connections, CDMA, 149
MAP (Mobile Application Part), GSM, 170
medical wireless applications, 338–339
MEO (Mid-Earth Orbit) satellites , 60, 69
messaging, SMS, 202
Metricom Ricochet, 259–264
micro-browsers, WAP, 199
micro/millimeter-wave radios, 52
microcell radios, Ricochet, 262
microcellular, 92
microwave communications, 12, 49
   bandwidth, 53–55
   CAPs, 51
   cellular, 50
   line of sight, 55
   multi-path fade, 55
   PCS, 50–51
   system installs, 50
   voice grade channels, 49
   water fade, 55
microwave repeater systems, 15
microwave systems, 12
Mobile Internet, 307–308, 311, 314
   handheld Internet devices, 313
   PDAs, 312
   Personal Companions, 313
   smart phones, 312
   UMTS operators, 315–316
   users, 314
   Web tablets, 313
Mobile IP, 244, 267
mobile surfers, WAP, 200
mobile to landline calls, 126
mobile units, GSM, 159
modems, 190–191
modulation, 9, 32
   amplitude, 34
   CDMA, 95
   digital, 36
   ETDMA, 94
   FDMA, 82–83
   frequency, 35
   signal envelope, 33
   TDMA, 84–85

Morse code, 4
Motorola
    Accompli, 258, 350
    Celestri, 64–65
MSAs (Metropolitan Service Areas), 113
MSCs (Mobile Switching Centers), 126
MSCs (Mobile Switching Centers), GSM, 160
MTP (Mobile Message Transfer Part), GSM, 166
MTSO (Mobile Telephone Switching Office), 114,
    119–120
multi-mode terminals, UMTS, 316
multi-path fade, microwave communications, 55
multiple access scheme, Teledesic, 277
multiplexing, 28
multiplied channel capacity, iDEN™, 175

N

NADC (North American Dual-mode Cellular), 305
Name Servers, Ricochet, 263
NAMPS (Narrowband AMPS), 24
NAMs (Numeric Assignment Modules), Over-the-
    Air Activation, 150
NetBlaster, 343
networks
    GSM, 160, 165, 170
    Teledesic, 278
NIFs (Network Interface Facilities), Ricochet, 262
NOCs (Network Operations Centers)
    LMDS, 294
    Ricochet, 262
noise, satellite systems, 69
non-LOS systems, fixed wireless, 337
notebooks, 257
NTT DoCoMo, wireless applications, 326

O

OFDM (Orthogonal Frequency Division
    Multiplexing), 245
Ohm's law, 42
orbital slots, 61
Over-the-Air Activation, 150
overhead channel, CDMA, 102

P

packet data services, CDMA, 152
packet efficiency, CDPD, 197
packet switching
    CDPD, 196
    Ricochet, 264
paging channel
    CDMA, 103
    GSM, 164
palmtops, 257
PANs (Personal Area Networks), 250, 342
parallel services, GSM and GPRS, 210
particles, electromagnetic radiation, 46
pass loss, radio frequencies, 48
PBSs (Personal Base Stations), cellular, 87
PCIA (Personal Communications Industry
    Association), 21
PCM (Pulse Code Modulation), 85, 139
PCS (Personal Communications Services), 25,
    50–51, 108, 134, 137
    CDMA, 142–148
    FDMA, 136
    make before break connections, 149
    market penetration, 140
    Over-the-Air Activation, 149
    providers, 152
    soft handoffs, 149
    TDD, 137–140
    TDMA, 136
    transmission methods, 138
PDAs (Personal Digital Assistants), 90, 256
    Bluetooth, 340
    wireless Internet, 312
PDC (Personal Digital Cellular), 137
person to person messaging, SMS, 202
Personal Companions, wireless Internet, 313
Phase 2+ GSM, 170
photons, 46
PHY layer, WLANs, 233–235
piconets, Bluetooth, 252
pilot channel, CDMA, 103
PLCP (Physical Layer Convergence Protocol), 234
PMD (Physical Medium Dependent), 234
Pocket Web, 311
PocketNet, 258
polarity, radio frequencies, 49

POPS (Points of Presence), 290
prisms, 46
propagation, 11–13, 48
protocols, WAP, 199
proxies, 266
pseudo-random code sequences, CDMA, 97
PSK (Phase Shift Keying), digital modulation, 36
PTM service, GPRS, 209
PTP service, GPRS, 209
PTTs (Post Telephone and Telegraph), 5

Q–R

QAM modulation, iDEN™, 175
R (resistance), 42
radio
  amplitude, 8
  carriers, 227
  channels, 25–36
  frequencies, 19–21, 40–49, 234
  HF band, 11
  microwave systems, 12
  modulation, 9
  propagation, 11
  resource management, GSM, 166
  satellite systems, 17–18
  short-wave, 8
  spectrum, 22, 187
  transmission types, 10
  UHF band, 12
  UWC-136 standard, 93
  VHF band, 12
radio-based links, WLANs, 223
radiotelephone calls, 13
random channels, GSM, 164
range of WLANs, 231
real-time videoconferencing, 346–347
Reed Solomon coding, CDPD, 197
registering frequencies, 7
regulatory bodies for spectrum usage, 22
reliability of WLANs, 231
repeaters, microwave systems, 15
retail wireless applications, 345
reusing frequency, 124
reverse channel, CDMA, 103
Ricochet, 259
  Gateways, 262
  microcell radios, 262
  Name Servers, 263
  NIFs, 262
  NOCs, 262
  packet switching, 264
  RF bands, 263
  WAPs, 262
  wireless modems, 260
RITL. *See* WLL.
RLP (Radio Link Protocol), CDMA, 150
roaming clients, WLANs, 237
RTS/CTS (request to send/clear to send),
    WLANs, 236

S

satellite communications, 56–57
  ACTS, 62
  bandwidth, 69
  C-band, 58
  DBS, 77
  downlinks, 57
  footprint, 57
  GEOs, 59, 64–65
  GPS, 77
  Ka-band, 58
  Ku-band, 58
  latency, 64, 68–69
  LEOs, 60, 70–72, 75–76
  links, 334–336
  markets, 65
  MEOs, 60
  noise, 69
  orbital slots, 61
  radio systems, 17–18
  security, 65
  Teledesic, 276
  transmission delay, 60
  transponders, 61
  UMTS, 317
  uplinks, 57
  VSATs, 62
satellite TV, 52
SCCP (Signal Connection Control Part),
    GSM, 166

SEA (spokesman election algorithm), 233
sectorized cell coverage, digital cellular
        systems, 123
security
    CDMA, 100
    satellite systems, 65
Semaphore, 4
service areas, iDEN™, 178
service providers, WLL, 286, 288
services
    GPRS support, 302
    GSM, 159
    spread spectrum, 99
    wireless Internet, 311
    WLANs, 224
SGSNs (Serving GPRS Support Nodes), 206
short and bursty transmissions, GPRS, 300
short-range radio technologies, 8, 252
signal envelope, modulation, 33
SIMs (Subscriber Identity Modules), GSM,
        159–160, 317
simultaneous voice/data transmission,
        CDMA, 151
Skybridge, 64
smart phones, 312, 350
smoke signaling, 4
SMR (Specialized Mobile Radio), 96
SMS (Short Message Service), 86, 140, 201–203
SNR (Signal-to-Noise ratio), 193
soft handoffs, CDMA, 149
software, wirelsss data, 214
SOHOware Net Blaster, 343
spectrum efficiency, iDEN™, 173
spectrum of frequencies, 7
spectrum regulation, 22, 187
speech coding formats, GSM, 165
speed of calls, iDEN™, 179
speed of light, 46
spread spectrum, 95
    CDMA, 142
    DS, 96
    FH, 96
    increased bandwidth, 97
    services, 99
    WLANs, 230
SS7 digital cellular systems, 130

standards groups
    frequency allocation, 19
    Internet, 270
STN-LCD phones, 331
subscriber base, GSM, 157
sync channel, CDMA 100, 103
system integrators, wireless data, 214

T

TACS (Total Access Control System), 24
TCP (Transmission Control Protocol), 271–272
    over Satellite, 268–269
    over wireless, 266–267
TDD (Time Division Duplex), 137
TDMA (Time Division Multiple Access), 29,
        84–87, 139–140, 164
telcos (local telephone providers), 5
Teledesic, 61, 64, 273
    fast packet switching, 275
    Ka-band, 276
    latency, 277
    multiple access scheme, 277
    network capacities, 278
    satellites, 276
    terminals, 273
telephone calls, 13
terminals, 158, 273
throughput
    CDPD, 197
    WLANs, 221, 231
TIA (Telecommunications Industry Association),
        frequency allocation, 19
tiered sites, digital cellular systems, 124
traffic channel, CDMA, 103–105
transmission
    CDMA, 101
    delay, satellite systems, 60
    methods, digital PCS, 138
    radio, 10
    wireless data, 188–189
    WLANs, 222
transponders, satellite systems, 61
troposcatter radio systems, 48
TTML (Tagged Text Markup Language), 258

U

UHF (Ultra High Frequency) band, 12
ultraviolet light, 47
UMTS (Universal Mobile Telecommunications
    Services), 25, 211, 298, 306–307, 310, 318
  3G, 305
  CDMA, 319
  IP compatibility, 318
  multi-mode terminals, 316
  satellite systems, 317
  USIMs, 317
  UTRA, 316
university wireless applications, 343–344
unlicensed spectrum, wireless data, 188
uplinks, satellite systems, 57
USIMs (UMTS Subscriber Identity Modules),
    UMTS, 317
USSD (unstructured supplementary services
    data), 199
UTRA (Universal Terrestrial Radio Access),
    309, 316
UTRAN, 212–213
UWB (Ultra-Wideband Radio), 253
UWC-136 HS standard, 93, 305
UWCC (Universal Wireless Communications
    Consortium), 168

V

VHF (Very High Frequency) band, 12
videoconferencing, 346-347
videophones, 259, 327
visible light, 46
VLR (Visiting Location Register)
  digital cellular systems, 127
  GSM, 161
vocoders, CDMA, 100, 144
VOD (Video on Demand), 331
VOD phones, 331
voice channels
  Bluetooth, 251
  microwave, 49
voice mail, SMS, 203
voice quality, TDMA cellular, 89
VoIP, 201, 308

VSAT (Very Small Aperture
    Terminal) satellites, 62
VSELP (Vector Sum Excited Linear Predictors)
    vocoder, 174–175

W

W-CDMA (Wideband CDMA), 210
WAP (Wireless Access Protocol), 199–201
WAPs (Wired Access Points), Ricochet, 262
WARC (World Administrative Radio
    Conference), 22
water fade, microwave communications, 55
wavelengths, frequency relationship, 46–47
WCAPs (Wireless CAPs), 52
Web applications, GPRS support, 302
Web tablets, 313
white light, 46
WI. *See* wireless Internet.
wireless applications, 324
  Alcatel, 325
  British Telecomm, 325
  CAGR (compound annual growth rate), 185
  carriers, 215
  CDPD, 195–197
  cellular modems, 191–192
  connections, 190
  CS-CDPD, 195
  data, 184–185, 216
  dial-up connections, 193
  EDGE, 92–93, 210
  GERAN, 211
  GPRS, 207–209
  Internet, 203–206
  location services, 351
  medical uses, 338–339
  microwave repeater systems, 15
  modem speeds, 190
  NTT DoCoMo, 326
  PDAs, 340
  retail, 345
  satellite links, 334
  spread spectrum, 97–99
  STN-LCD phones, 331
  videoconferencing, 346–347
  VODphones, 331

wristwatch phones, 332
radio, 3–12, 187
radiotelephone calls, 13
satellite radio systems, 17–18
SMS, 201–203
SNR ratio, 193
software, 214
spectrum regulation, 187
system integrators, 214
telephone calls, 13
transmission, 188–189
UMTS, 211
unlicensed spectrum, 188
UTRAN, 212–213
WAP, 199–201
WITs, 186
wireless downstream systems, 280–283
Wireless Internet, 203, 326–329
    future services, 204–205
    GPRS, 206–209
    MM (Mobility Management), 204
wireless LAN adapters, WLANs, 228
wireless modems, Ricochet, 260
wireless network middleware, proxies, 266
wireless standards
    cdma2000, 319–320
    EDGE, 298, 304–305
    GPRS, 298–301
    mobile Internet, 307–308
    UMTS, 298, 306–307, 316–319
    VoIP, 308
wireless systems
    5-UP, 247
    802.11a standard, 244–245
    access points, 241
    Bluetooth, 248–252
    BRAN, 247
    HiperLAN/2, 247
    HomeRF, 238
    infrared, 37
    Mobile IP, 244
    mobility, 243
    open standards, 248
    short-range radio technologies, 252
    wireless adapters, 242

wireless videophones, 259
WITs (Wireless Intelligent Terminals), wireless
    data, 186
WLANs (Wireless LANs), 220
    802.11 specifications, 232–235
    access points, 227
    ACK, 236
    applications, 224
    bandwidth speeds, 221
    configurations, 228–229
    contention window, 235
    cost benefits, 226
    coverage, 231
    CRC, 238
    CSMA/CA, 234
    infrared, 223, 234
    integrity, 231
    market share, 224
    PLCP, 234
    PMD, 234
    radio carriers, 227
    radio frequency transmissions, 234
    radio-based links, 223
    range, 231
    reliability, 231
    roaming clients, 237
    RTS/CTS, 236
    spread spectrum, 230
    throughput, 221, 231
    transmissions, 222
    wireless LAN adapters, 228
WLL (Wireless Local Loop), 173, 279–280, 285
    bandwidth, 289
    service providers, 286–288
wristwatch phones, 332

X–Z

X-rays, 47
X.25, GPRS support, 301
Z (Impedance), 42

Contact Us for Training

WWW.TCIC.COM

Custom Training

Instructor-Led Training

Keynote Speaking

Consulting

Books

Computer-Based Training

1-800-322-2202
info@tcic.com